Deep Learning Generalization

This book provides a comprehensive exploration of generalization in deep learning, focusing on both theoretical foundations and practical strategies. It delves deeply into how machine learning models, particularly deep neural networks, achieve robust performance on unseen data. Key topics include balancing model complexity, addressing overfitting and underfitting, and understanding modern phenomena such as the double descent curve and implicit regularization.

The book offers a holistic perspective by addressing the four critical components of model training: data, model architecture, objective functions, and optimization processes. It combines mathematical rigor with hands-on guidance, introducing practical implementation techniques using PyTorch to bridge the gap between theory and real-world applications. For instance, the book highlights how regularized deep learning models not only achieve better predictive performance but also assume a more compact and efficient parameter space. Structured to accommodate a progressive learning curve, the content spans foundational concepts like statistical learning theory to advanced topics like Neural Tangent Kernels and overparameterization paradoxes.

By synthesizing classical and modern views of generalization, the book equips readers to develop a nuanced understanding of key concepts while mastering practical applications.

For academics, the book serves as a definitive resource to solidify theoretical knowledge and explore cutting-edge research directions. For industry professionals, it provides actionable insights to enhance model performance systematically. Whether you're a beginner seeking foundational understanding or a practitioner exploring advanced methodologies, this book offers an indispensable guide to achieving robust generalization in deep learning.

Liu Peng is currently an Assistant Professor of Quantitative Finance at the Singapore Management University (SMU). His research interests include generalization in deep learning, sparse estimation, and Bayesian optimization.

Deep Learning Generalization

Theoretical Foundations and Practical Strategies

Liu Peng

CRC Press
Taylor & Francis Group
Boca Raton London New York

CRC Press is an imprint of the
Taylor & Francis Group, an **Informa** business

A CHAPMAN & HALL BOOK

First edition published 2026
by CRC Press
2385 NW Executive Center Drive, Suite 320, Boca Raton FL 33431

and by CRC Press
4 Park Square, Milton Park, Abingdon, Oxon, OX14 4RN

CRC Press is an imprint of Taylor & Francis Group, LLC

© 2026 Liu Peng

First edition published by Routledge 2025

ISBN: 978-1-032-84190-8 (hbk)
ISBN: 978-1-032-84189-2 (pbk)
ISBN: 978-1-003-51160-1 (ebk)

DOI: 10.1201/9781003511601

Typeset in CMR10
by KnowledgeWorks Global Ltd.

For my wife Zheng and children Jiaxin, Jiaran, and Jiayu.

Contents

1

Unveiling Generalization in Deep Learning

Most of us strive to excel in academic exams. From a young age, we are often encouraged to study diligently, not just for practice tests but also for the final exam. In the realm of machine learning, this parallels the process of training a model on a dataset where both inputs (questions) and outputs (answers) are known, much like a set of practice exam questions with their solutions. Our objective is to train the model on these "practice exam" data and then apply it to unseen data, representing the final exam.

The overarching goal is to develop a robust model that not only performs well on the training data (the practice exam) but also generalizes effectively to the test data (the final exam). In deep learning, this entails uncovering patterns that are not merely memorized from the training set but are broadly applicable to new, unseen data. If the model is overfitting, memorizing the practice exam questions without grasping the underlying concepts and patterns, it can perform poorly on the final exam. In contrast, a well-generalized model is better equipped to handle new questions (test data) confidently, reflecting a deeper understanding of the fundamental patterns of the data.

Depending on the model's ability to generalize, we can encounter four possible outcomes: good performance on both the training and test data, representing strong generalization because the model captures underlying patterns without overfitting; poor performance on both, reflecting underfitting when the model fails to learn meaningful relationships; strong performance on the training set yet poor performance on the test set, indicating overfitting caused by memorizing training examples rather than understanding broader patterns; and, in some rare cases, unexpectedly better performance on the test data than on the training data, which can occur due to randomness or a fortuitous alignment with the test distribution.

Figure 1.1 summarizes these results, where the lower left quadrant represents the worst-case scenario: poor results on both the training set and the test set. Although unfortunate, this outcome is somewhat predictable: insufficient learning from the training data is likely to result in poor generalization to unseen test data. The key is to fully leverage the training data, which comes with labeled answers, to develop a model that captures meaningful patterns that can be applied to future data. If the model fails to learn adequately from the training set, it is highly unlikely to perform well on new data.

Occasionally, we might observe good performance on the test set alongside poor results on the training set, a situation depicted in the upper left

DOI: 10.1201/9781003511601-1

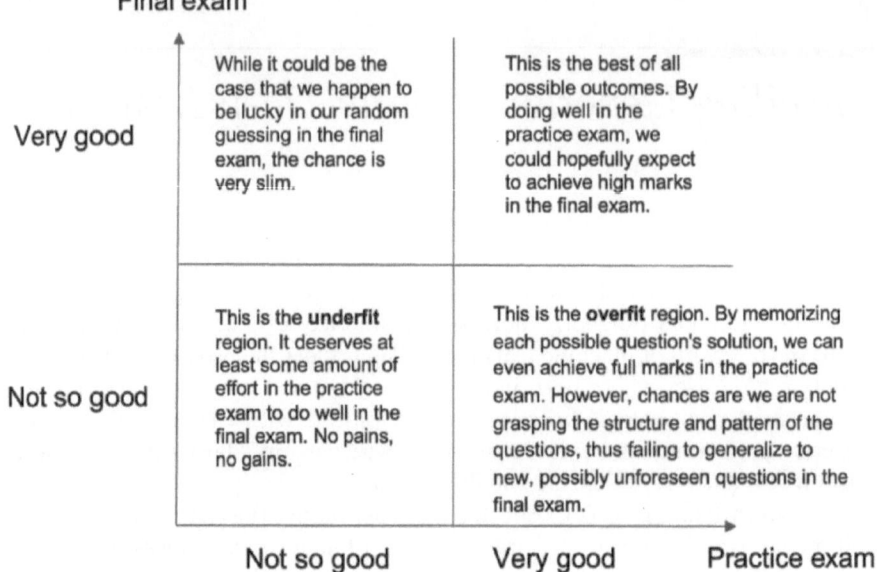

FIGURE 1.1
There are four different outcomes for the practice and final exams, where the practice exam contains question-answer pairs and the final exam only contains the questions. This also corresponds to the four possible scenarios in model development, where the available training data contain the input-output pairs and the test data only contain the input data. The trained model is assessed on both training and test data, although a more significant focus is given to the latter.

quadrant. Although it may seem promising at first glance, this outcome is both uncommon and often attributable to random chance rather than to a thorough understanding of the underlying patterns. In practice, such luck rarely translates into stable generalization and does not typically guarantee strong performance on future tasks.

The upper right quadrant represents the ideal outcome: the model excels in both the training set and the test set. In this scenario, the model has uncovered the essential patterns in the training data and applied them successfully to new, unseen data. Achieving this balance usually involves carefully refining the model during training so that it neither underfits nor overfits, thereby enabling effective generalization to future tasks. This quadrant, where performance aligns well with the underlying goal of robust learning, captures the essence of strong generalization. In fact, the model may even produce a lower test set error and thus generalize better if it overfits the training data to some extent, as evidenced by the double-descent phenomenon in deep neural networks (more on this in later chapters).

The lower right quadrant highlights the phenomenon of overfitting, where the model, much like a student overly fixated on practice exam scores, has been tuned too extensively to the training set. In this case, it memorizes specific answers rather than internalizing the broader concepts underlying those answers. Although the model shows strong performance on the training data, it struggles when faced with unfamiliar questions on the test set. This mirrors a student who relies on rote memorization for practice exam questions, only to falter when confronted with differently phrased or more challenging problems in the final exam.

In deep learning, overfitting arises when a model becomes overly complex and is tuned too closely to the training set, effectively memorizing its details rather than grasping the underlying structures. Because the final exam (test set) inevitably differs from the training data, performance deteriorates when broader patterns are not properly learned. In contrast, underfitting occurs when the model is too simple to capture even the essential features in the training set, leading to poor results on both the training and test data. Among these two extremes, overfitting is more prevalent in deep learning, where the pressure to optimize for immediate gains can overshadow the ultimate goal of solid generalization. The key lesson is not to memorize the training answers, but instead to learn the underlying patterns that can be applied to new, unseen data.

In this context, the lower right quadrant, representing overfitting, shows a model that excels on the training set but struggles with new unseen data. This issue can be mitigated through regularization, a collection of strategies — such as weight loss (L2 regularization), dropout, or early stopping — aimed at preventing the model from simply memorizing the training set. These slight adjustments help guide the model toward the broader underlying structures in the data rather than focusing solely on specific examples. Thus, a properly regularized model is more likely to generalize effectively, yielding accurate predictions on unseen data in the test set.

Generalization stands as the foremost objective of any predictive model, where the ability to perform effectively on unseen data becomes the paramount concern. In deep learning, regularization is essential to mitigate overfitting, offering a variety of methods that control the complexity of the model and reduce the discrepancy between training and test results. Especially when training data are scarce or model capacity is limited, these techniques become indispensable to ensure that the model retains its ability to learn key patterns without simply memorizing the training examples. Approaches to regularization can involve adjusting the data, the model architecture, the cost function, or the optimization routine, all with the aim of balancing the learning algorithm to achieve robust generalization.

Many practical deep learning projects face a significant hurdle in transitioning from the lower right quadrant, where overfitting dominates, to the upper right quadrant, where robust generalization prevails. This challenge raises several important questions: how can overfitting be identified in the first place, and at what point does it become a serious concern? What methods are available to mitigate its impact, and why do some strategies yield more success than others? In the sections that follow, we will delve into these issues, discussing not only how overfitting can be detected but also how it can be systematically addressed to achieve durable, effective performance on unseen data.

1.1   Introducing Generalization

A typical machine-learning model is defined by the number of parameters it contains and by how its architecture directs the interplay between these parameters and the input features to generate predictions. Deep neural networks, in particular, are renowned for their complexity because they usually contain an extensive number of parameters and employ multilayered architectures capable of capturing intricate, highly nonlinear relationships. This heightened complexity can make them remarkably powerful, yet it also renders them more susceptible to overfitting, wherein the model not only learns the genuine patterns but also memorizes random noise in the training data, ultimately degrading performance on unseen test data.

In practical settings, although a subset of the data might be collected with exceptional precision and carry little to no noise, most real-world data contains some degree of randomness or measurement error. If a model starts internalizing these noisy artifacts, it risks mistaking them for genuinely important trends. Although this approach can yield high accuracy on the training data, the same idiosyncratic noise typically does not recur in the test data, causing the model to underperform when confronted with new examples where such noise may not be present or assume another form. In our running exam metaphor, this scenario parallels a student memorizing incorrect answers from the practice exam; when those exact inaccuracies do not appear on the final, the student's performance inevitably suffers.

The central objective in model development is thus not to force the model into a flawless fit for the training set, but rather to capture the genuine patterns that remain consistent between the training and test data. In the same way that we would not blindly trust every practice exam response, we should train our models with a measure of caution, seeking to build solid inferential capabilities that fit any particular set of examples in the unknown test set.

Interestingly, deliberately introducing noise can sometimes enhance the learning process. This strategy, referred to as data augmentation, involves injecting small controlled variations into input features, model parameters, or even target labels. When carefully applied, data augmentation fosters robustness by teaching the model to handle diverse data variations, reducing its susceptibility to overfitting on idiosyncratic noise. It mirrors studying

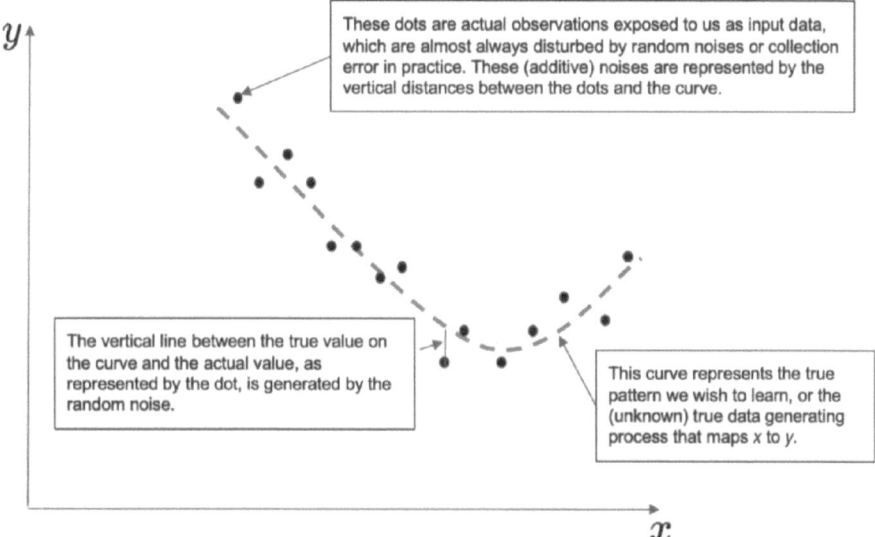

FIGURE 1.2
The true relationship we want to learn, as represented by the curve, and the training data available for model development, as represented by the dots. The (vertical) deviations between the curve and the dots are due to random observational noises (assumed to be additive in this case) or collection errors, which almost always appear in practice. Our goal is to use these dots to build a model to approximate the true pattern as much as possible, namely the real relationship between input x and output y. At the same time, the learned model needs to be robust enough to avoid distractions from the noises.

multiple slightly altered versions of similar questions or answers, accompanied by potential typos, to prepare for an exam: by encountering these variations, you develop greater adaptability and are better equipped to deal with any unexpected twists in the final exam.

In fact, if the training set encompasses every possible variation found in the test set, even a relatively simple method like k-nearest neighbors can achieve near-perfect generalization since effectively every question in the final exam has already been seen in practice. However, in real-world scenarios, such perfect overlap is exceedingly rare. This is where noise injection and data augmentation become invaluable, introducing controlled variations that encourage models to handle unexpected deviations in new data. By making them more resilient against randomness and overfitting, these techniques can potentially foster robust performance on unseen data and ultimately help us build models that generalize effectively.

Let us visualize the true pattern, the noise, and the model in action. In Figure 1.2, the black dots represent the actual observations, or input-output pairs visualized in a two-dimensional coordinate system, while the dotted line shows the true underlying function that we want our model to learn. However, in practice, these observations rarely align perfectly with the true function due to random noise or errors in the data collection process. This noise leads to vertical deviations between the dots and the curve, making it more challenging to approximate the true function. Our task is to build a model that approximates the dotted line as accurately as possible.

Although striving for solid generalization is crucial, reaching that goal can be complicated by two frequent pitfalls mentioned before: underfitting and overfitting. A more complex model, such as a neural network, often provides a richer framework for capturing nuanced patterns than a simpler model like linear regression. However, if no mechanisms are in place to check the complexity of the model—such as through regularization, careful architecture design, or other constraints—the model may become overly sensitive to noise in the data. This overfitting scenario emerges when the model not only learns the genuine underlying relationships but also memorizes random fluctuations, ultimately impairing its ability to generalize to unseen test data.

Conversely, if we select a model that is too simple, for example, using a straight line (a linear regression model) where in fact a more flexible curve is a better fit, it may lack the necessary capacity to capture the real complexity in the data, resulting in underfitting. In this situation, the model is overly constrained and fails to learn the crucial structures underlying the data, leading to suboptimal performance on both the training set and likely any unseen test data.

There are two common scenarios where underfitting can occur. The first happens when we restrict ourselves to a simple linear regression model, which struggles to capture more complex, nonlinear relationships due to insufficient model complexity. In the left panel of Figure 1.3, the straight line represents the model, which is not flexible enough to approximate the true underlying function, shown as a curved line. This lack of flexibility leads to underfitting, where the model cannot capture the complexity of the data. However, when the amount of training data is severely limited, as illustrated in the middle panel, a complex model may not have sufficient information to learn effectively. Faced with this data scarcity, a simpler model, like linear regression, often performs better because it is less prone to overfitting. Conversely, a more complex model, such as a neural network, is likely to memorize the small number of available examples, undermining its ability to generalize. In this situation, choosing a simpler model becomes the safer bet; however, the core issue lies more with the lack of data than with the model itself. Even the most sophisticated algorithm cannot perform well when trained on only a handful of examples, as it becomes overly sensitive to noise.

In each of these scenarios, the model, illustrated by the straight line, remains too simple to encapsulate the underlying complexity of the actual pattern, as depicted by the curved line. Whether due to insufficient flexibility (as

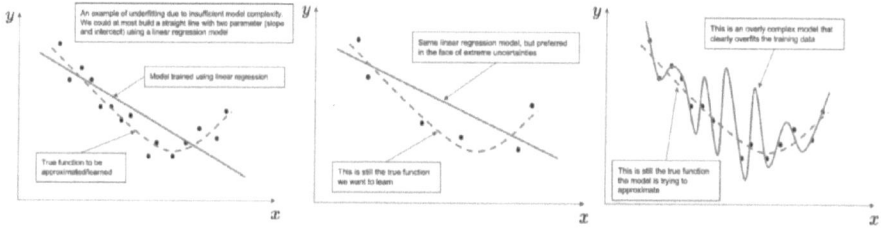

FIGURE 1.3
(Left) An example of fitting a nonlinear function using a linear function, which
underfits the training data. The model needs to be flexible enough to be able
to learn the pattern of the true function, which is often highly complex in
reality. (Middle) Another underfitting example, where the same linear regression model is now a preferred choice due to extremely limited training data.
When only exposed to very limited training data, it is totally fine to use linear regression, which tends to generalize better than a complex model such as
neural networks. Deep neural networks tend to be highly complex and are thus
prone to overfitting. However, the problem in this case is more on the training
data and less on the model choice; we are limited by what we can do in the
first place. Any model built on top of six observation points alone will likely
be sensitive and not robust to the noises in the data. (Right) An example of
overfitting due to an excessively complex model. The model perfectly fits all
dots, giving a zero error rate in the training set. However, the wiggly curve
comes at a price of generalizability — the model's predictions for new input
data will be far off from the truth. In other words, it fails to generalize to
future unseen data in the test set. The excessive complexity leads to a large
generalization gap.

in the first scenario) or insufficient data (as in the second), the model fails to
adequately represent the richer structure present in the data.

When a model becomes excessively complex, as illustrated in the right
panel of Figure 1.3, the result is often overfitting—a situation in which the
model contorts itself to accommodate every point in the training set. This
leads to a highly irregular, wiggly curve that may align perfectly with the
training data but fails to perform well on new, unseen inputs. The gap between
training performance and test performance (loosely understood as the "generalization gap") becomes substantial because the model has effectively learned
the noise rather than the legitimate structure in the data. Consequently, it
performs poorly on the test set, missing the true underlying pattern in favor
of fitting idiosyncratic details from the training set.

Regularization techniques are frequently employed to combat both underfitting and overfitting by steering the model toward an appropriate level
of complexity. They work by imposing penalties for unnecessary complexity, thereby narrowing the generalization gap and preventing the model from
merely memorizing training noise. In line with Occam's razor—where the

simplest explanation often proves the most robust—simpler models tend to exhibit better generalization. Even deep neural networks, known for their capacity to achieve extremely low training error, benefit from regularization strategies such as weight decay (L2 regularization), dropout, or early stopping. By favoring a less convoluted model over a more complex one, we strike the right balance between flexibility and resilience. Since the ideal level of complexity cannot always be predicted in advance, regularization provides a mechanism for adjusting the model during training to ensure it captures the true underlying patterns while remaining impervious to noise.

Striking the right balance in model complexity is thus a crucial consideration. When the model is overly simple, it struggles to capture the actual curvature in the data, leading to underfitting. In contrast, if it is too complex, it may produce a highly fluctuating curve that conforms to noise rather than reflecting the genuine patterns, leading to overfitting. The ultimate aim is to adopt a model flexible enough to reveal the underlying relationships in the data yet sufficiently robust to disregard the random noise.

1.2   The Curse of Dimensionality

Modern machine-learning models often become over-parameterized and prone to overfitting due to the rapid expansion of Big Data. In recent years, the "four Vs" of Big Data—volume (the quantity of data), velocity (the speed at which data is generated and collected), variety (the different types of data), and veracity (the uncertainty or noise in the data)—have all grown substantially. This transformation stems from advances in data collection methods, such as live-streaming video, web activity tracking, cheaper computational resources, easily accessible machine-learning frameworks, and the fast pace of business transactions. Although these developments open up remarkable opportunities, they also pose formidable challenges for both data storage and modeling. As data becomes more abundant, the incentive to design larger and more complex models to handle it grows, making them increasingly susceptible to overfitting if left unchecked.

One immediate obstacle lies in the high dimensionality of modern datasets. The abundance of features, or dimensions, in Big Data imposes a "curse of dimensionality," making it increasingly challenging to detect and interpret meaningful patterns. While traditional methods like linear regression perform well with smaller, more manageable feature sets, they often struggle when confronted with highly dimensional inputs. For example, image data can contain thousands of features, necessitating more sophisticated algorithms to effectively extract, identify, and transform the critical information needed for accurate predictions. As shown in Figure 1.4, image-based applications highlight this complexity and underscore the need for more powerful techniques that can handle extensive feature spaces.

FIGURE 1.4
This image taken in Singapore's Central Business District (CBD) illustrates the high dimensionality of typical color images. The shape of the image is (400, 225, 3), where 400 is the width, 225 is the height, and 3 refers to the red, green, and blue color channels. In total, this single image contains 270,000 features (400 × 225 × 3).

Although images can be quite intricate, text data can be just as, if not more, complex. A single document may contain numerous words whose meanings shift according to context, forcing the model to discern which words matter most and how they interact. This can quickly escalate to tens of thousands of features. Fortunately, regularization techniques help address the challenges posed by high dimensionality. One approach, known as sparse learning, targets a limited but important subset of features while discarding the rest, effectively assuming that only a small number of features are truly predictive. The irrelevant features are assigned zero-valued coefficients and effectively removed. A well-known example of sparse learning is the Least Absolute Shrinkage and Selection Operator (Lasso), which imposes a penalty on the magnitude of coefficients, reducing overfitting and enhancing generalization by automatically performing feature selection.

As an example of sparse learning, imagine we want to classify the image in Figure 1.4 as either a country or a city scene. Pixel features that depict tall buildings or other urban structures would likely be more predictive than those capturing grass, sky, or other less distinguishing elements. By enforcing sparsity constraints, the model can assign zero coefficients to pixels that contribute little or no information, effectively silencing these irrelevant features and fo-cusing its attention on the most salient predictors. This targeted emphasis helps improve both the interpretability and performance of the classification task.

Regularization offers multiple intuitive viewpoints here. From a geometric perspective, introducing a penalty restricts how large the estimated coeffi-cients can grow, effectively nudging some of them toward zero. In a Bayesian framework, this can be seen as adopting a prior that assigns higher probability to smaller coefficients, favoring those that are often zero. From the standpoint of robust optimization, methods like Lasso serve to shield the model from the effects of noise, yielding more stable coefficients that are less susceptible to random fluctuations in the data.

1.3   More on Underfitting and Overfitting

Underfitting occurs when the effective model complex is insufficient compared to the available training data, causing it to underperform on both the training and test sets. In contrast, overfitting emerges when a model appears highly accurate in potentially noisy training set but suffers on unseen data. In this overfitting scenario, the generalization error increases, reflecting a substantial gap between the model's performance in the known training set (empirical risk) and its performance in the unknown test set (true risk). The overarch-ing concern here is generalizability: we ultimately want models that not only match their training data but also produce reliable predictions on data they have never encountered before.

Returning to our exam analogy, the effort we devote to studying can be viewed as a stand-in for model complexity: the harder we study, the more nuanced our mental model becomes. As illustrated in Figure 1.5, when we begin to increase our study hours and sharpen our problem-solving skills, both our practice exam (training) and final exam (test) scores show tangible improvements. However, the closer we get to perfection, the more substan-tial the effort required to achieve even marginal gains. Progressing from a score of 60 to 70, for example, is generally easier than moving from 80 to 90, and each step closer to 100 demands exponentially greater investment. If we push our study to the point of attaining flawless results on the practice exam, we run the risk of overfitting—remembering the specific questions and answers rather than honing general problem-solving abilities. This approach may achieve near-perfect scores on the practice set but does little (due to mere memorization) to ensure strong performance on the final exam. In fact, the narrower the focus on memorizing precise answers, the larger the gap be-tween our practice and test scores, reflecting a widening generalization gap. Moreover, given the abundance of practice questions, there is no guarantee that every paper is noise-free or without mistakes, thus making an overfitting approach even more dangerous.

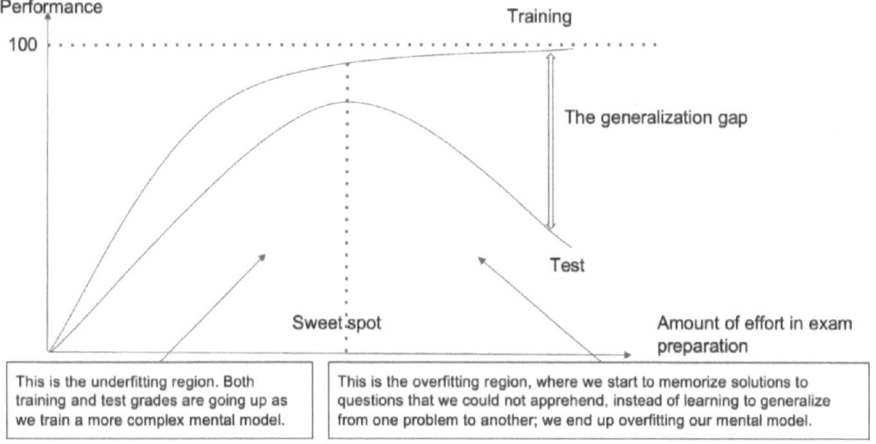

FIGURE 1.5
The evolution of training and test performances as more effort is invested in exam preparation. We start in the underfitting region, where both training and test performances improve as the mental problem-solving model becomes stronger. However, the model starts to overfit after passing the sweet spot, where test performance begins to decline, even though training performance continues to improve with added complexity.

The essential objective is to develop a mental model that excels on both the practice and final exams. A well-generalizing model achieves an optimal level of complexity: it is sufficiently sophisticated to recognize and capture the core problem-solving patterns, yet not so intricate that it fails to apply those insights to unfamiliar questions. Consequently, minimizing the generalization gap—ensuring that performance on unseen data nearly matches performance on known data—becomes paramount.

That brings us to the concept of a "sweet spot": an optimal balance in which performance on the final exam is maximized. In this range, the model (or mental framework) is complex enough to capture the essential patterns, yet not so intricate that it merely memorizes the practice set and overlooks general principles. Beyond this sweet spot, increasing model complexity (or study effort) becomes counterproductive, leading to overfitting. In technical terms, the sweet spot marks the point at which test error is minimized and generalization capability is at its highest, reflecting an ideal alignment of model capacity and data complexity.

An optimal model steers clear of both underfitting and overfitting. It maintains sufficient flexibility to capture the true patterns in the training data while remaining simple enough to avoid being misled by noise. By achieving this delicate balance, it achieves strong performance on previously unseen data that follow the same data-generating distribution, successfully generalizing beyond the confines of the training set.

1.4   Understanding Bias-Variance Trade-off

The concepts of underfitting and overfitting, illustrated by the sweet spot in Figure 1.5, are intimately linked to the bias-variance trade-off, a principle that applies broadly to machine-learning tasks like regression, classification, and reinforcement learning. In a regression context, bias refers to the degree to which, on average, the model predictions align with the true target values, linking to the idea of an expectation. High bias occurs when the model systematically misses the mark, suggesting that it fails to capture the key patterns in the data and consistently generates off-target predictions. Typically, a model that is too simple has high bias, leading to underfitting. Returning to our exam analogy, a simplistic mental model translates into poor scores on both practice and final exams, reflecting an inability to grasp and apply essential problem-solving patterns.

On the other hand, variance indicates how much a model's predictions change in response to small variations in the training data. High variance typically signals overfitting, as the model focuses too heavily on perfectly matching the training samples, making its predictions overly sensitive and unstable when presented with new data. Ideally, we aim for low variance, where predictions remain consistent despite minor fluctuations in the training set. In practice, a high-variance model may achieve strong performance on the training data by memorizing it, yet it often underperforms on the test set. This aligns with our exam analogy: precisely memorizing the practice exam answers can lead to a perfect score on the practice set, but it generally fails in the final exam if any of those memorized answers are incorrect or if the exam questions differ even slightly. Such significant variations in the predictions from one set of data to another (high variance) indicate that the model has overfitted the training data.

In machine learning, both bias and variance represent key sources of error that we strive to minimize. Yet reducing both simultaneously is inherently difficult. Lowering bias typically entails increasing the model's complexity, which in turn can elevate variance and the likelihood of overfitting. Conversely, curbing variance usually involves simplifying the model, potentially raising bias and causing underfitting. This tension between addressing bias and variance is referred to as the bias-variance trade-off.

Referring to Figure 1.6, we see four possible scenarios that describe varying levels of bias and variance. In the first region, the model exhibits high bias and low variance, which typically indicates that the model is too simplistic and underfits the data. Moving to the second region, both bias and variance are

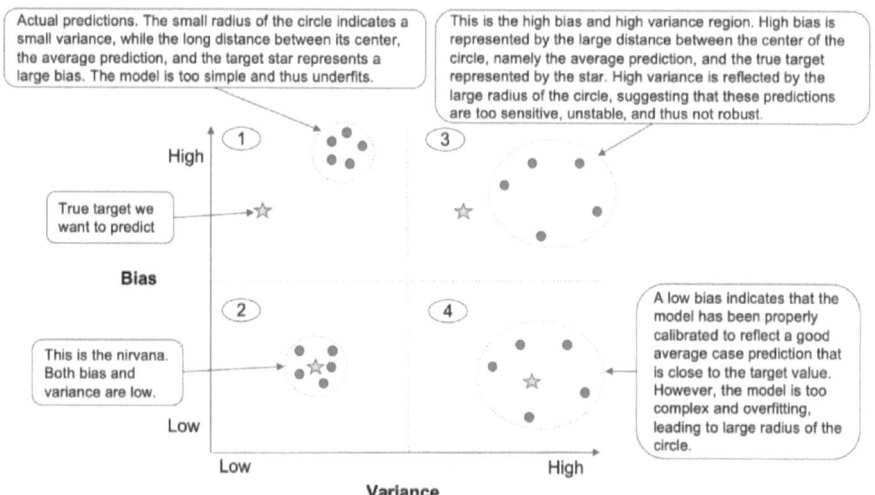

FIGURE 1.6
Four possible bias-variance scenarios. The first region has high bias and low variance due to the trained model being too simple and possibly underfitting. The second region is the ideal state, with bias and variance both being low. The third region has high bias and variance, suggesting the model deserves further training to learn the underlying pattern. The fourth region indicates an overfitting model that has low bias but high variance.

low, representing the ideal scenario in which the model captures the essential patterns of the data without overfitting or underfitting. The third region shows high bias and high variance, suggesting that the model is still in the early stages of training and has not yet internalized the fundamental relationships in the data; as a result, it does not perform well in either the training or test sets. Finally, the fourth region features low bias but high variance, signaling that the model has latched onto the training data too closely—memorizing not only the true patterns but also the noise—thus exhibiting clear signs of overfitting.

In practice, training often begins in the third region, where both bias and variance remain high. As the model is refined and becomes more intricate, we progress toward the first region, where bias is reduced and the model's predictions gain stability. Ideally, we would like to arrive at the second region, characterized by low bias and low variance—a setting in which the model balances accuracy on the training data with robust performance on unseen data. However, it is more common to observe models that keep increasing in complexity and overshoot into the fourth region, where overfitting sets in and variance climbs too high.

Effective regularization methods can help steer a model toward achieving the ideal combination of low bias and low variance—the sweet spot for

Deep Learning Generalization

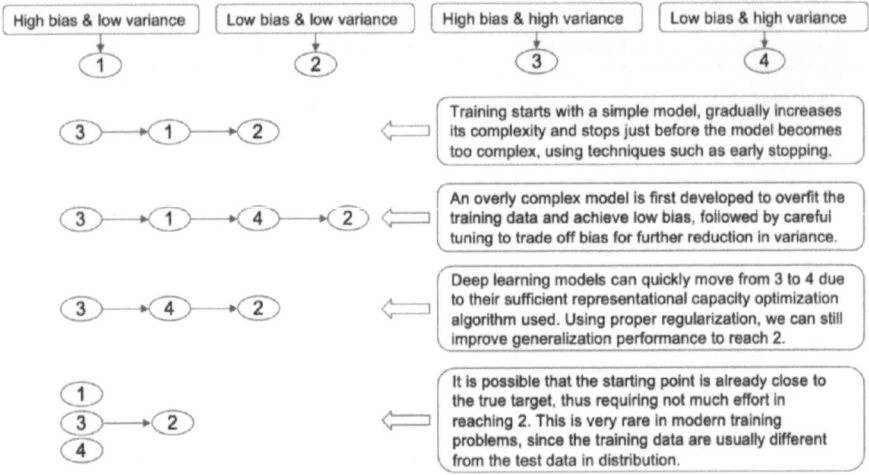

FIGURE 1.7

Example model training paths using proper regularization. There are multiple ways to build a generalizable model with low bias and variance, as in region 2. Depending on the similarity between the training set and the test set, as well as the model class and training strategy used, the model training path could travel in multiple ways. With proper regularization, the final trained model will hopefully reach to region 2.

generalization. As shown in Figure 1.7, there are various training trajectories that can guide a model to this optimal region, depending on the specific regularization strategies and training procedures employed. For example, one path might move the model from region 3 (high bias, high variance) to region 1 (high bias, low variance), and then proceed to region 2 (low bias, low variance). Another path might take the model through region 4 (low bias, high variance) before it ultimately settles in region 2. Each path demonstrates a different route to achieving a well-generalized model, shaped by factors like the match between training and test data, the chosen model architecture, and the specific training regimen. For instance, some models might more naturally move from region 3 (high bias, high variance) to region 1 (high bias, low variance) before arriving at region 2, while others might briefly pass through region 4 (low bias, high variance) and then settle at region 2. Regardless of the path, the overarching goal is to guide the model to region 2, where accuracy (low bias) and stability (low variance) are both maintained, ensuring robust performance on unseen data.

Achieving the right balance entails choosing a model that is neither overly simplistic nor needlessly complex, presuming sufficient data exists to capture the problem's inherent variability. Concretely, this means steering clear of underfitting as well as overfitting. The central objective is to drive down the

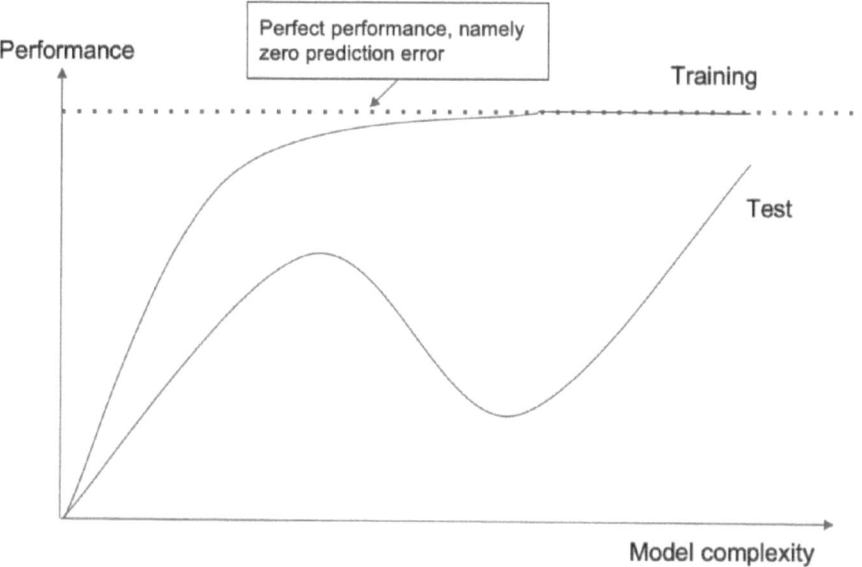

FIGURE 1.8
The double ascent phenomenon using complex models such as deep neural networks. As the model continues to become more complex, we will start to observe increasing performance in the test data, where the variance continues to decrease when the bias is zero.

total error—encompassing both bias and variance—through a careful balancing act that reduces each source of error without allowing one to dominate. By managing this trade-off effectively, we can attain lower total error and ultimately enhance the model's performance. By thoughtfully managing variance through regularization, the model gains stability and robustness, reducing its likelihood of overfitting and enhancing its capacity to generalize to unseen data. While this process may introduce a slight increase in bias, the model's average predictions will still hover near the true target values, preserving accuracy without sacrificing stability.

It is worth mentioning that the classic bias-variance trade-off does not universally hold. In highly parameterized models such as deep neural networks, with sufficient training, variance can diminish even while bias remains close to zero. This phenomenon is often referred to as "double descent," (or double ascent), as depicted in Figure 1.8. Initially, increasing model complexity can cause test performance to drop because of overfitting. However, beyond a certain point, further increases in complexity actually improve test performance again, accompanied by a reduction in variance—even though bias stays low. This resurgence in performance characterizes the essence of the double-descent phenomenon.

In this regime, the usual bias-variance analysis does not necessarily capture a clear trade-off between the two. When working with deep neural networks, sufficient training time and iterations can allow the model to move beyond the overfitting phase and ultimately achieve stronger generalization on the test data. This raises the question: why do deep learning models, even when they appear to overfit the training set, still end up generalizing effectively to new data? The answer seems to lie in the very nature of deep neural networks and, in particular, their training via stochastic gradient descent (SGD), as discussed in [15]. Through a process sometimes referred to as implicit regularization, SGD discourages pure memorization of the training examples, enabling the model to hone in on patterns that extend to unseen data and thus attain better test performance.

1.5   More on the Model Training Path

Let's delve further into the model training trajectory. The standard machine-learning procedure begins with minimizing a particular loss function—often computed as the discrepancy between the model's predictions and the actual labels in the training set—with the intent that this model will also perform effectively on unseen test data. This approach, widely known as empirical risk minimization, parallels our exam analogy: by practicing on known questions (the training set), we aim for strong performance on the final exam (the test set). However, this process can yield four distinct outcomes, each capturing how thoroughly the model has learned from its training data and how successfully it generalizes beyond it.

In the first scenario, both the training set error and test set error are high, indicating that the model is underfitting. It struggles to extract meaningful patterns from the training data, either because it is too simple or because it has not been trained sufficiently to grasp the underlying structure. To address this issue, one can increase the model's complexity, extend the training duration, refine the hyperparameters, or even switch to a more suitable class of models.

The second scenario, in which we see high training set error yet low test set error, can occur but is extremely uncommon. Under usual assumptions where the training and test sets come from the same data-generating process, a model that underperforms on the training set is unlikely to excel on the test set. While random chance or particular quirks in the data might occasionally lead to better performance on the test set, the conventional expectation is that struggling to fit the training data implies difficulty in handling new data as well.

In the third scenario, we observe low error on the training set combined with high error on the test set—a hallmark of overfitting. The model, having become excessively complex, almost memorizes the training examples and

FIGURE 1.9

Four different scenarios at the end of training. We wish to build a model that not only achieves low training error but also, and more importantly, low test error.

consequently struggles to handle unseen data. While it may perform near-flawlessly on the training set, it is overly sensitive to noise or inconsequential patterns, leading to poor test performance. This common challenge is usually mitigated through various regularization strategies that encourage the model to learn generalizable patterns instead of intricately fitting every training instance.

Lastly, the fourth scenario—considered the ideal case—occurs when both the training set error and the test set error are low. This outcome suggests that the model has effectively captured the crucial patterns of the data, avoiding both underfitting and overfitting, and demonstrating strong generalization to unseen data.

Figure 1.9 illustrates these four scenarios, helping us visualize the different outcomes that can result from the model training process.

Let us change our perspective. Our main objective is to pick a model—out of potentially many—that performs exceptionally well on both the training set and the test set. Each model corresponds to a specific set of parameters and a configuration of mathematical operations—like addition, multiplication,

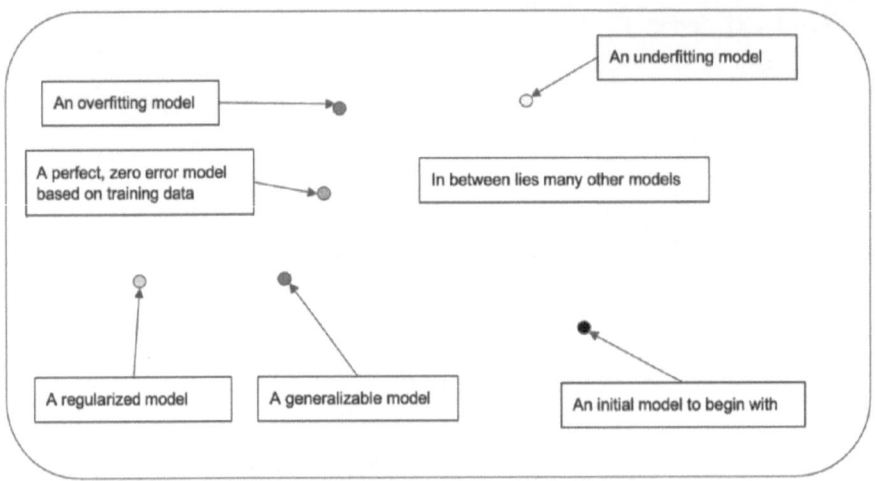

FIGURE 1.10
The total space of a (big enough) model class. Each point represents a particular model, and the relative positioning is arbitrary in this example. Our goal
is to develop a properly regularized model where both training error and test
error are low.

and other transformations—that govern how information flows and interacts
within the model. All these possible configurations, taken together, constitute
what we call a model class. If we consider a neural network, for instance, its
model class includes every conceivable arrangement of weights and biases in
the prespecified architecture. Our challenge is to navigate this vast collection
of candidate models and identify the configuration that yields strong performance on the training data while still generalizing effectively to unseen data.

What we aim to do is locate one of the "good" models within this vast class
of models, one that can generalize effectively to new unseen data, by carefully
training it on the available training set. As depicted in Figure 1.10, this model
class, which can be extremely large for complex architectures such as neural
networks, encompasses numerous possible parameter configurations. Some of
these configurations will extend well to the test set, while others may overfit
or underfit, failing to capture the right balance of complexity and robustness.
Ultimately, our task is to train the model so that we settle on a configuration
that achieves not only strong performance on the training set but also reliable
accuracy when confronted with the test set, thus ensuring good generalization.

The model training process in Figure 1.10 can also be thought of as moving
through different model configurations in the search for the one that best
fulfills a chosen optimization criterion. Given the breadth of potential paths,
the outcome hinges on factors like the chosen model class, training algorithm,
and specific training regimen. Figure 1.11 offers a concrete illustration of this
path in the case of a classification and regression tree (CART) model. Initially,

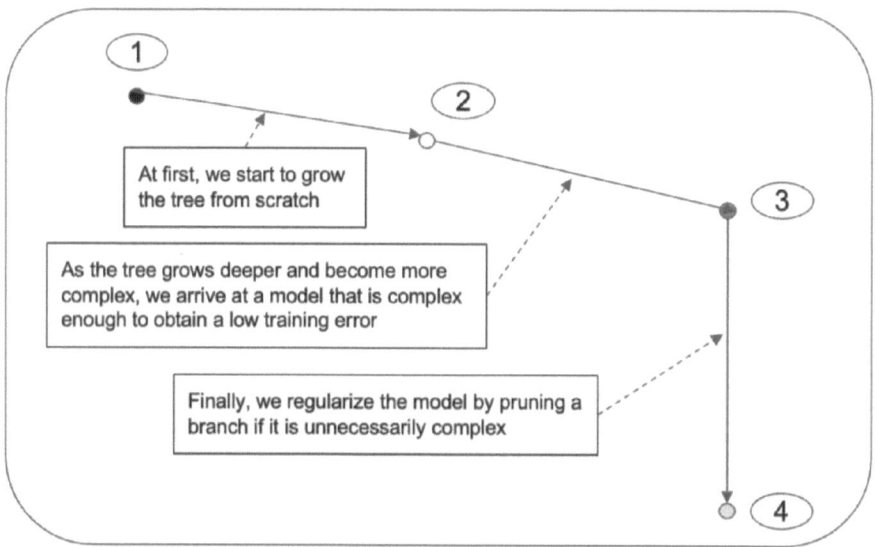

FIGURE 1.11
Example path from training a random forest model. Depending on the model class and training strategy used, the specific training path may vary for different models.

the tree underfits (from the first point to the second), owing to a lack of complexity. As more branches and leaves are added, the model grows overly complex, risking overfitting (the third point). By pruning unneeded branches, one can effectively regularize the model, paring it back to a simpler, more generalizable structure that avoids overfitting and performs reliably on unseen data.

In practice, it is often the case that more than one model configuration can generalize effectively to the test set. As illustrated by the dashed circle in Figure 1.12, these configurations form a collection of models that achieve strong performance on unseen data. Our primary objective is to train the model so that it ends up inside this dashed circle, where generalization is secure. When overfitting becomes an issue, regularization techniques step in to constrain the model complexity, thereby increasing the likelihood of converging to one of the well-generalizing solutions within this collection.

In a typical training routine, the objective is to minimize training error, often referred to as empirical risk. This metric indicates how well a model might generalize, given that training and test sets follow similar distributions. Unfortunately, in many real-world scenarios, this assumption may not strictly hold. Consequently, minimizing empirical risk alone might not guarantee robust generalization. By incorporating regularization into the training process, an approach known as structural risk minimization, practitioners can better

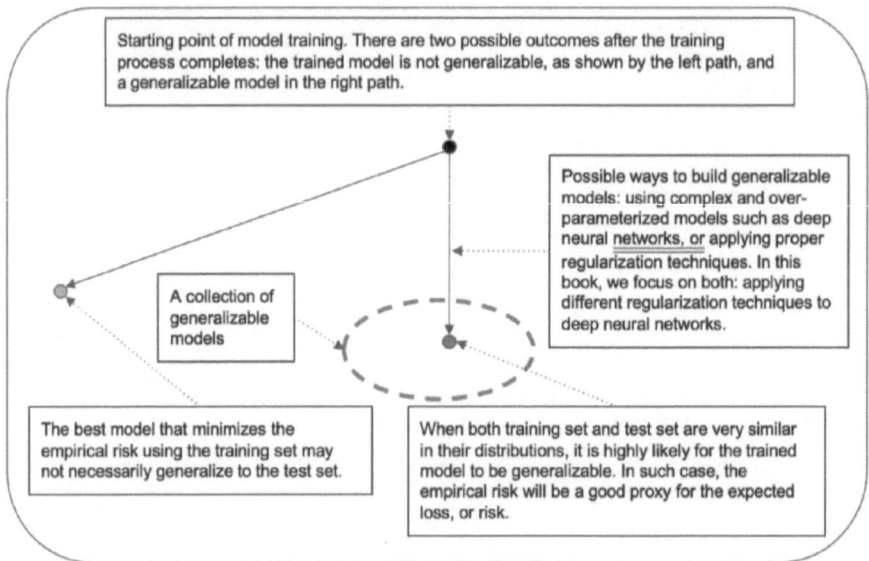

FIGURE 1.12
Two possible outcomes of training a model in terms of its generalizability to
the test set.

control model complexity and improve performance on the test set, even when
there is a slight mismatch between training and test distributions.

To elaborate, a model that achieves near-perfect predictions on the training data is not guaranteed to perform equally well on unseen test data. This
situation arises because, during training, the model is exposed to a particular range of input-output pairs and learns how to map those inputs to their
corresponding outputs. That ability to transfer learned knowledge to new inputs hinges on whether they resemble the patterns observed in training. Consequently, if the training set and the test set originate from similar distributions, a well-trained model will likely exhibit strong generalization. However,
this similarity cannot always be taken for granted, especially when random
noise, limited data, or other complexities disrupt the representativeness of the
training set in real-world scenarios.

To mitigate the generalization gap, one strategy involves deploying deep
neural networks trained with SGD, leveraging the "double-descent" phenomenon as mentioned earlier. Although deep networks are characteristically
high-dimensional, SGD capitalizes on this complexity by implicitly regularizing the model, often leading to strong performance on both training and
test sets. In parallel, explicit regularization techniques—such as weight decay, dropout, or early stopping—can also be applied to directly rein in model
complexity, further reducing the risk of overfitting and strengthening generalization.

Note that different models trace distinct routes through the search space during training, and some are more prone than others to converge on a well-generalizing solution (those located in the dashed circle in Figure 1.12). For example, when deep neural networks, often extremely complex, are adequately trained, they can achieve a training error that is near zero. Nevertheless, they can still perform well on the test set, which at first seems contradictory because highly complex models are usually expected to overfit and fail to generalize. As introduced earlier, this apparent paradox is resolved by the double-descent phenomenon, a hallmark behavior in over-parameterized models like deep neural networks.

In deep neural networks, once the model exceeds a certain complexity threshold, it can exhibit a decrease in variance even after its bias effectively reaches zero as training continues, a key observation based on the double-descent phenomenon. Furthermore, because the true function to be approximated is often highly complex and nonconvex, the loss landscape may exhibit multiple local minima. Each of these local minima can correspond to an equally effective model, capable of achieving zero training error or minimal regularized loss. Consequently, it is not uncommon to end up with multiple high-performing models trained on the same data, all of which provide a comparably strong generalization in practice.

These "best" models can collectively be visualized as a set of optimal solutions. As shown in Figure 1.13, a Venn diagram depicts three overlapping circles, each representing a set of models defined by distinct performance criteria. Circle 1 encompasses models that perfectly fit the training set and achieve zero training error. Circle 2 includes models refined by regularization, which manages complexity and mitigates overfitting. Finally, Circle 3 captures models that excel in generalizing, evidenced by their low error in the test set. The intersection of these circles represents a collection of models that not only fit the training data but also handle noise effectively and demonstrate strong performance on unseen data.

The key takeaway is that any pair of these circles can intersect, yet their overlap is never guaranteed. For instance, just because a model fits the training data perfectly and achieves zero training error (Circle 1) does not imply that it will also generalize successfully to unseen data (Circle 3). Likewise, employing a particular form of regularization (Circle 2) does not automatically ensure that the model will end up generalizing well. Each circle thus represents a different aspect of model performance, and their intersections show that while some models can meet multiple criteria—fitting the training set, being regularized, and generalizing effectively—others might only fulfill one or two of these requirements.

When we use highly over-parameterized models such as deep neural networks in conjunction with suitable regularization methods, the odds of finding a model that generalizes effectively increase considerably. In other words, we become more likely to land at the intersection of all three circles: models that fit the training data perfectly (achieving zero training error) are regularized to control complexity and ultimately generalize well on the test set.

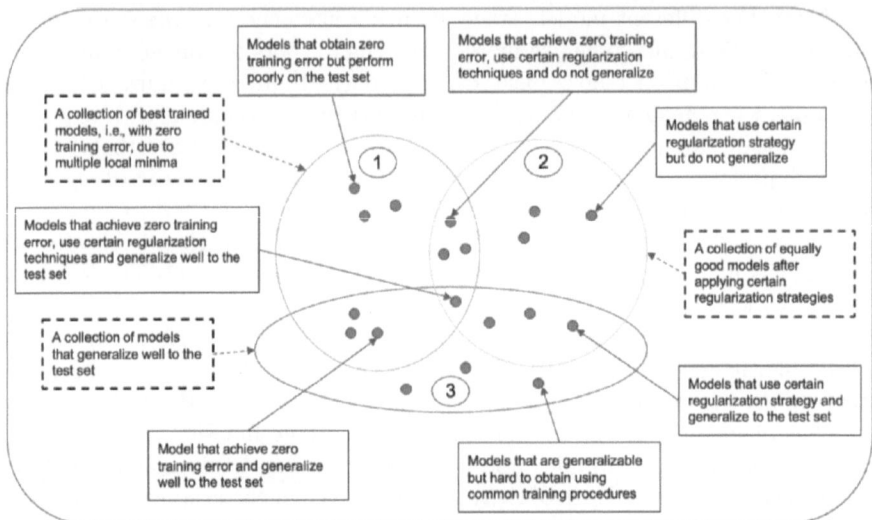

FIGURE 1.13

Three different collections of candidate models. It should be noted that either reaching zero error in the training set, or applying regularization, or a combination of both, does not guarantee building a generalizable model in the end. What happens is that using over-parameterized models like deep neural networks, as well as applying proper regularization techniques, tends to end up with a model that generalizes well to the test set.

1.6   Understanding the Model Training Process

A typical model training process begins with a set of training data. In supervised learning tasks, this data is made up of pairs of inputs and outputs, where each input entry includes multiple features that describe the observation from various perspectives. The associated output represents the true target value, serving as the "correct answer" that guides the training process. By repeatedly evaluating how well the model's predictions match these target values, the training algorithm refines the mapping function so that it can accurately transform any given input into its corresponding output.

A trained model is composed of two key elements: the parameters, which can be modified during training, and the architecture, which governs how those parameters interact with the input data to generate the predicted output. The model prediction is then compared to the actual target value, and the difference between the two, called the error, serves as a performance measure. This error reveals how close the predicted output of the model is to the true target, indicating whether the model reliably captures the underlying data patterns or if further adjustments are needed.

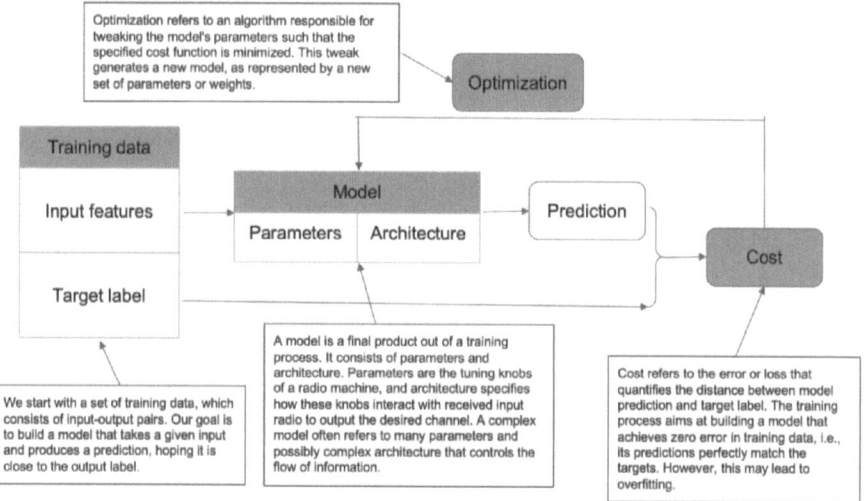

FIGURE 1.14

Example of a typical model training process. The workflow starts with the available training data and gradually tunes a model. The tuning process requires matching the model prediction to the target output, where the gap is measured by a particular cost function and used as feedback for the next round of tuning. Each tuning produces a new model, and we want to look for a model that minimizes the cost.

In the training process, an optimization algorithm systematically updates the model's parameters (and optionally its architecture) with the goal of minimizing the error. After each update, the model makes a new prediction, and the error is recalculated, creating a feedback loop that continually refines the model's performance. This process repeats until the error is reduced to a satisfactory level, indicating that the model has effectively learned from the data.

The entire training process, depicted in Figure 1.14, features four key components: training data, model, cost, and optimization that collectively shape how a model learns. Each of these components can be a strategic point for introducing regularization to boost generalization performance. A notable example is data augmentation, a technique often used in object detection tasks to enhance the training data itself. By systematically injecting noise or applying transformation, such as shearing, rotation, or flipping, to the input images, we create a larger, more varied dataset. These newly augmented images are then included in the original training set, allowing the model to encounter a wider range of visual scenarios. As a result, it learns to handle variations more effectively, reducing overfitting and improving robustness to unseen data.

This diversity in training data plays a vital role in enhancing the ability of the model to generalize to new unseen inputs. By exposing the model to a broader range of variations during training, we effectively prepare it to handle the many ways objects or scenes can appear in the real world. This explains why data augmentation is especially advantageous in tasks such as image classification, where the model must recognize objects across different viewing angles, lighting conditions, or distortions.

1.7   An Overview of Different Regularization Techniques

Many regularization methods have been devised that target various aspects of the machine-learning pipeline: data, model, cost, optimization, or a hybrid of these. In practice, the effectiveness of each technique depends on the specific problem at hand, with some methods delivering more immediate performance gains than others. However, much of the current research on these approaches exists in a fragmented form, typically treating each technique in isolation. To address this issue, our aim is to consolidate popular regularization strategies within a single cohesive framework.

Building on the previously discussed training process, training a neural network follows a cycle of feeding data into the network architecture, allowing the model parameters to interact with those inputs to produce predictions, computing the error, and then adjusting the model's weights based on that error. The top panel of Figure 1.15 shows how various regularization techniques can be grouped according to these stages in the model training pipeline, reflecting a model-centric view of how different methods fit into the broader workflow.

Alongside the model-centric categorization, regularization methods can also be viewed through the lens of implicit versus explicit techniques. Implicit regularization influences the model's behavior without directly imposing additional penalty terms or constraints. A key example is SGD, a widely used optimization algorithm in deep learning. During each iteration of SGD, the model is updated using a small batch of examples randomly selected from the training set, thus introducing variability (random gradient) into the training trajectory. This process immerses the model in diverse data patterns, indirectly enhancing its robustness. Moreover, because the parameter updates are somewhat uncorrelated, SGD inherently acts as a form of regularization, steering the model toward solutions that generalize better on the test set.

On the other hand, explicit regularization involves directly managing certain aspects of model training, often by constraining or penalizing excessive model complexity. The middle panel of Figure 1.15 illustrates several commonly used implicit and explicit regularization strategies. Among these, dropout is noteworthy because it can act as an implicit and explicit regularizer: by randomly "dropping" certain neurons during training, it explicitly reduces the capacity of the network at each iteration, while also implicitly introducing variability that improves the model's ability to generalize.

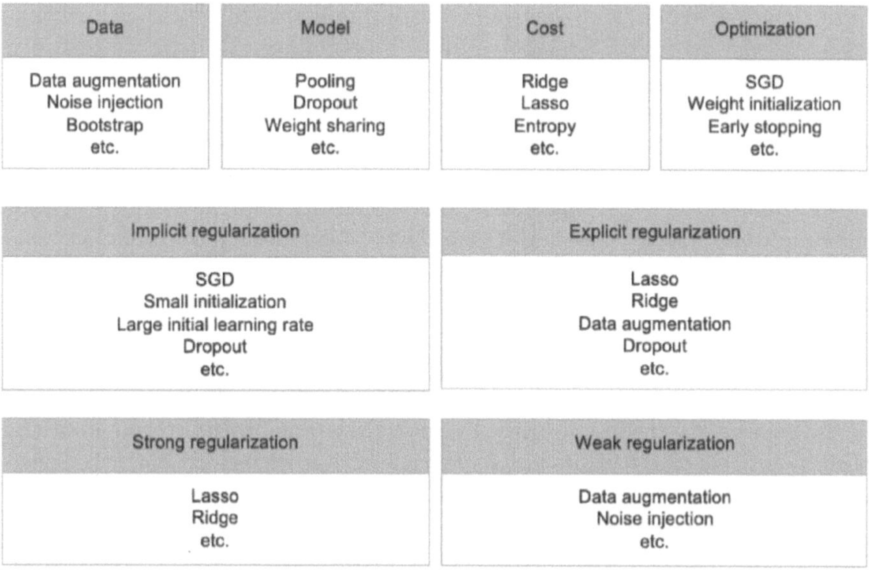

Data	Model	Cost	Optimization
Data augmentation Noise injection Bootstrap etc.	Pooling Dropout Weight sharing etc.	Ridge Lasso Entropy etc.	SGD Weight initialization Early stopping etc.

Implicit regularization	Explicit regularization
SGD Small initialization Large initial learning rate Dropout etc.	Lasso Ridge Data augmentation Dropout etc.

Strong regularization	Weak regularization
Lasso Ridge etc.	Data augmentation Noise injection etc.

FIGURE 1.15
Different ways of categorizing regularization techniques.

Finally, in the bottom panel of Figure 1.15, regularization techniques are classified as either strong or weak, reflecting the degree of their impact on the model. For instance, explicit methods such as Lasso have a strong regularizing effect because they directly prune model complexity by forcing certain coefficients to zero, effectively discarding unimportant features. In contrast, methods such as data augmentation work by providing the model with multiple versions of the same dataset, enabling it to learn robust patterns from variations in the data. This results in a more gradual or indirect influence on generalization, making data augmentation a comparatively weaker regularizer.

For instance, with image data, data augmentation can reflect geometric properties such as symmetry, rotation, or permutation. These properties are often hard for the model to learn directly from the raw images. By augmenting the data with these transformations, the model is more likely to capture these properties, improving its ability to generalize to new, unseen images.

1.8 Toward Model Generalization via Hyperparameter Tuning

In addition to the implicit and explicit regularization methods discussed, another powerful way to boost model generalization is hyperparameter tuning.

The primary goal here is to determine the optimal set of hyperparameters that maximize the model performance on new unseen data, ultimately improving its ability to generalize. Hyperparameter tuning also provides fine-grained control over different parts of the training pipeline, such as the model's architecture, learning rate, and regularization strength, allowing us to navigate the balance between underfitting and overfitting more effectively. By systematically refining these hyperparameters, we help ensure that the model is neither too simple nor too complex, a critical factor for robust generalization. One commonly used technique for hyperparameter tuning is Bayesian optimization, which reasons about the next set of hyperparameters to try out in a principled manner [9].

It is important to note that hyperparameters are parameters that the model does not optimize during training; rather, they are chosen (or pre-fixed) by the user before training begins. Because they remain fixed throughout the learning process, their values can have a significant impact on both the speed and quality of the model's convergence. Common examples of hyperparameters include the learning rate, which dictates the step size for gradient-based updates; the number of hidden layers in a neural network, determining its depth and representational capacity; and the kernel size in a support vector machine (SVM), which influences how the model classifies data by shaping the feature space.

For example, in a neural network, adjusting the learning rate helps to strike the best compromise between fast convergence and a stable learning trajectory: higher rates can speed up training but may cause instability, while lower rates lead to more reliable but slower improvement. Likewise, tuning the number of hidden layers or the number of neurons shapes the architecture's capacity to capture the data's complexity without making the model excessively large or unwieldy. Meanwhile, refining the strength of regularization (another hyperparameter in weighting the importance of the regularization term compared with the loss term) can prevent the model from overfitting to the training data, helping it maintain the flexibility needed to generalize successfully to unseen examples.

In general, the path to achieve good generalization differs based on the characteristics of the dataset and the specific problem domain, which means that there is no panacea that works best in all settings. Different techniques tend to excel under different circumstances, influenced by factors such as the complexity of the model, the amount of data available, and the nature of the underlying task.

1.9   Summary

In this chapter, we have examined how the notion of "training for an exam" in machine learning helps illuminate the core challenge of building models that perform well not only on training data (the practice exam) but also on unseen data (the final exam). We introduced four possible outcomes, ranging from underfitting (weak performance in both training and test sets) to strong generalization (good performance on both), along with two more extreme scenarios of rare or unbalanced performance. Central to this discussion is the idea that models should learn the underlying patterns of the data rather than merely memorizing examples that may contain noise or inaccuracies. In deep learning, overfitting is of particular concern because neural networks, with their large number of parameters, can easily learn spurious details that reduce their ability to generalize. Nevertheless, we discussed that well-structured training procedures, coupled with suitable regularization, can bridge the gap from overfitting to robust generalization. The ultimate goal is to strike a balance—sometimes called the "sweet spot"—where the model has enough capacity to capture meaningful relationships but not so much that it memorizes irrelevant details.

We then shifted focus to how over-parameterized models, such as deep neural networks, can eventually achieve excellent test performance despite appearing to overfit at first. This behavior, known as the "double-descent" phenomenon, occurs when further increases in model complexity (such as a number of parameters or training epochs), coupled with sufficient training, can actually reduce test error again after an initial surge in overfitting. In simpler contexts, the bias-variance trade-off explains why increasing model capacity often lowers bias but raises variance, while reducing capacity does the opposite. However, deep neural networks trained using SGD can exhibit an implicit regularization effect, meaning that even in high-dimensional parameter spaces, they are nudged toward solutions that generalize more reliably than one might expect. Alongside these implicit effects, explicit regularization techniques (e.g., dropout, L2 weight penalties, early stopping) are empirically useful tools in preventing the network from locking onto unhelpful details. This chapter emphasized that the capacity of modern deep learning models, though immense, can still be harnessed to build robust predictors with the right design of architecture, training strategies, and regularization.

Another major theme was the "curse of dimensionality," in which real-world data with numerous features, such as high-resolution images or large vocabularies in text, require models powerful enough to extract useful relationships. Traditional methods (like linear regression) often underfit when faced with such complexity, prompting the need for deeper architectures and more sophisticated feature processing. However, simply increasing the model size invites a higher risk of overfitting if not properly regularized. Techniques like data augmentation introduce controlled perturbations, such as shearing, rotation, flipping, or noise injection, so that the model learns robust, general features and is less susceptible to random fluctuations. Sparse learning methods, such as Lasso, tackle dimensionality by forcing small or unimportant coefficients to zero, thus improving interpretability and performance. Throughout these discussions, we showed how explicit (direct) regularization and implicit (indirect) regularization both play critical roles in shaping the behavior of a model.

Finally, we considered the broader workflow of model selection, training, and hyperparameter tuning. Every model lies in a large "model class" of possible parameter configurations. By minimizing empirical risk (training error), we hope to discover a region of parameter space that also generalizes well. However, real-world complexities, such as noisy measurements or mismatches between training and test distributions, make this process nontrivial. Proper regularization, structural risk minimization, and careful hyperparameter choices, such as learning rate, network depth, or regularization strength, help us navigate to reliable solutions. We saw that multiple "good" models may emerge (e.g., those achieving zero training error, those with strong regularization, and those that generalize well), and the intersection of these sets is where we want to end up. In the end, there is no single recipe for guaranteed success: data augmentation, bias-variance management, double-descent insights, and hyperparameter tuning must be tailored to each problem's data characteristics. Nonetheless, when applied judiciously, these strategies unify to produce models that not only excel on their "practice exams" but also confidently tackle the "final exam" of unseen, real-world data.

2

Introduction to Statistical Learning Theory

Continuing from our running example in Chapter 1, many of us strive to excel in school exams. From a young age, we are encouraged not only to prepare thoroughly for practice tests but also to ensure that we can perform well on the final exam. In the context of machine learning, this mindset is closely mirrored by the training process. Specifically, we train a model on a labeled dataset, where the inputs $\mathbf{x}$ (the "questions") and the outputs y (the "answers") are known, just as we study and refine our understanding using practice tests. Ultimately, however, our goal is to deploy or evaluate the trained model on previously unseen data, which we can think of as the final exam. In doing so, we capture the critical concept of generalization: just as true academic success is gauged by performance on the final exam, the success of a model is determined by how effectively it performs on data beyond the training set.

To formalize, we begin with a training dataset $\{(\mathbf{x}_i, y_i)\}_{i=1}^n$, where each pair $(\mathbf{x}_i, y_i)$ consists of an input $\mathbf{x}_i$ and its associated label y_i. Here, $\mathbf{x}_i \in \mathcal{X}$ lies in the input space $\mathcal{X}$, often referred to as the space of input instances (or observations). This space encompasses all possible representations of the objects we intend to model, in either classification (modeling a categorical y) or regression setting (modeling a continuous y). For example, in an image classification setting, each $\mathbf{x}_i$ might be a feature vector encoding properties such as color intensity, texture, or shape extracted from an image. By representing these features as vectors, we create a structured way to capture the essential characteristics of the data for further processing in our learning algorithms.

The label y_i resides in the label space $\mathcal{Y}$, which encompasses all possible categories or classes. In a classification setting, $\mathcal{Y}$ is typically a finite set of discrete labels, such as $\{\text{cat}, \text{dog}, \text{cow}, \dots\}$. Each y_i thus identifies one of these categories, specifying the true class of the corresponding input $\mathbf{x}_i$. Our principal objective in classification is to map every instance in $\mathcal{X}$ to the correct category in $\mathcal{Y}$ as much as possible, thereby minimizing classification errors. However, this error measure is nonconvex, making direct minimization challenging. Instead, we often employ a more tractable loss function, such as the cross-entropy loss, to guide the learning process effectively.

Thus, given the training data $\{(\mathbf{x}_i, y_i)\}_{i=1}^n$, our task is to learn a mapping or function $f : \mathcal{X} \to \mathcal{Y}$ that can accurately assign labels to previously unseen instances. Concretely, we seek a function f such that, for any new input $\mathbf{x}$, the predicted label $f(\mathbf{x})$ coincides with the true label y as frequently as possible. This goal amounts to reducing the number of elements in $\mathcal{X}$ for which $f(\mathbf{x})$

DOI: 10.1201/9781003511601-2

diverges from the true label, effectively minimizing the classification error across the input space.

The function f that accomplishes this mapping from $\mathcal{X}$ to $\mathcal{Y}$ is commonly referred to as a classifier. We assess the effectiveness of a classifier by examining how well it generalizes, meaning its ability to predict labels accurately for new inputs not included in the original training set. In other words, good generalization implies that the learned mapping is not merely memorizing the training examples but capturing underlying patterns that extend to unseen data. In practice, we often quantify a classifier's performance using its classification error, which is defined as the fraction of instances in $\mathcal{X}$ that are assigned incorrect labels. This error rate can be estimated using a held-out test set, cross-validation procedures, or other validation strategies designed to evaluate the model's performance on data beyond the training set.

As discussed in Chapter 1, depending on how well a model generalizes, we may observe four distinct outcomes. One possibility is strong generalization, where the model achieves high accuracy on both the training and test datasets. Another possibility is underfitting, in which the model exhibits poor performance on both sets, indicating it has not captured the essential patterns in the data. A third scenario is overfitting, characterized by excellent performance on the training data but poor results on unseen test data. Finally, there is the rare occurrence of an anomalous generalization, where the model performs poorly on the training data but unexpectedly excels on the test data, an outcome often attributed to random chance or an unusual alignment with the test distribution.

Formally, we denote by $\mathcal{L}_{\text{train}}(\theta)$ the training loss and by $\mathcal{L}_{\text{test}}(\theta)$ the test loss, where θ represents the parameters of the model. The training loss is caused by

$$\mathcal{L}_{\text{train}}(\theta) = \frac{1}{n} \sum_{i=1}^{n} L\big(y_i, f(\mathbf{x}_i; \theta)\big),$$

and the test loss is

$$\mathcal{L}_{\text{test}}(\theta) = \mathbb{E}_{(\mathbf{x},y) \sim \mathcal{D}_{\text{test}}} \big[L\big(y, f(\mathbf{x}; \theta)\big)\big],$$

where L is a chosen loss function, such as mean squared error for regression or cross-entropy for classification, and $\mathcal{D}_{\text{test}}$ is the distribution of test data. Intuitively, the training loss measures how well the model fits the labeled examples in the training set, whereas the test loss estimates how well the model generalizes to new, unseen data.

The training loss $\mathcal{L}_{\text{train}}(\theta)$, computed by averaging the chosen loss function L across all training examples, reflects the degree to which the predictions of a model match the labels in the training dataset. For example, in regression problems, a common choice for L is the squared error for the i-th instance, defined as,

$$L\big(y_i, f(\mathbf{x}_i; \theta)\big) = \big(y_i - f(\mathbf{x}_i; \theta)\big)^2,$$

which measures the discrepancy between the predicted value $f(\mathbf{x}_i; \theta)$ and the true value y_i.

In contrast, classification tasks often employ the cross-entropy loss, which quantifies how dissimilar the predicted probability distribution is from the true label distribution. Specifically, denoting

$$\hat{p}_{i,c} = \frac{e^{f_c(\mathbf{x}_i; \theta)}}{\sum_{k=1}^{K} e^{f_k(\mathbf{x}_i; \theta)}}$$

as the predicted probability for class c of the i-th instance, the cross-entropy loss for a single training example $(\mathbf{x}_i, y_i)$ can be written as

$$L\big(y_i, f(\mathbf{x}_i; \theta)\big) = -\sum_{c=1}^{K} \mathbf{1}(y_i = c) \log \hat{p}_{i,c}.$$

Here, $\mathbf{1}(y_i = c)$ is an indicator function that is 1 if $y_i = c$ and 0 otherwise. This ensures that only the log probability of the correct class contributes to the loss, thus encouraging the model to assign a high probability to the true label and align its prediction to the real target.

In this chapter, we delve into the mathematical framework of statistical learning theory and discuss its key principles, laying the groundwork for subsequent topics in machine learning.

2.1 Introducing Generalization

A typical machine-learning model is defined by its parameters θ and by the way its architecture regulates the interaction between these parameters and the input features to generate predictions. Deep neural networks, in particular, often have a large number of parameters as well as intricate designs, which make them highly expressive yet susceptible to overfitting. In such cases, the model may memorize both the true patterns in the training data and the random noise, causing it to perform poorly when presented with new, unseen data. Understanding and mitigating this risk is a central theme in developing effective learning systems, where the goal is to strike a balance between model complexity and generalization performance.

Our primary objective in learning is to minimize the expected risk, also referred to as the true risk $\mathcal{L}_{\text{test}}(\theta)$ from earlier,

$$R(\theta) = \mathbb{E}_{(\mathbf{x},y)\sim\mathcal{D}}\big[L\big(y, f(\mathbf{x}; \theta)\big)\big],$$

where $\mathcal{D}$ represents the true, yet typically unknown, data-generating distribution from which the input-output pair $(\mathbf{x}, y)$ is sampled. The quantity $R(\theta)$ thus captures the expected average loss throughout the data distribution, serving as the most widely used measure of the predictive performance of any model.

In practice, because $\mathcal{D}$ is not accessible, we instead work with the available training data and minimize the following empirical risk (referred to as $\mathcal{L}_{\text{train}}(\theta)$ earlier),

$$R_{\text{emp}}(\theta) = \frac{1}{n} \sum_{i=1}^{n} L\big(y_i, f(\mathbf{x}_i; \theta)\big),$$

which approximates the expected risk by averaging the loss over the finite training dataset. However, simply minimizing $R_{\text{emp}}(\theta)$ does not guarantee a reduction in $R(\theta)$, as this is related to how good the model class is in approximating the true risk (also called the Bayes risk) and how good the training procedure is in obtaining the best risk within the model class (more on this later). In particular, if the model becomes too tuned to the training data, capturing noise rather than true patterns can occur, thus reducing the generalizability of the model in new unseen samples.

2.1.1 Revisiting underfitting and overfitting

Consider the underlying true function $f^*(\mathbf{x})$ that we are looking to learn. However, due to the presence of the random noise term ϵ, assumed to be additive, the observed outputs y are generated according to:

$$y = f^*(\mathbf{x}) + \epsilon,$$

where ϵ is commonly assumed to be drawn from a normal distribution, $\epsilon \sim \mathcal{N}(0, \sigma^2)$, with a constant variance. Our objective is to find an estimate $\hat{f}(\mathbf{x})$ that approximates $f^*(\mathbf{x})$ as accurately as possible while not fitting too much to the noise term. However, when the model is overly flexible, it may capture both the underlying function and the random noise, which we already know as overfitting. This results in poor generalization to new data because the model has effectively learned to replicate noise rather than the true signal.

When it comes to underfitting, it means that the model lacks the capacity to capture the true structure in $f^*(\mathbf{x})$. One reason might be inadequate model complexity; for instance, employing a simple linear model $y = \theta_0 + \theta_1 x$ to capture data that, in reality, follows a nonlinear relationship $y = \theta_0 + \theta_1 x + \theta_2 x^2$. In such scenarios, the model exhibits high bias (high distance between $\mathbb{E}[\hat{f}(\mathbf{x})]$ and $\mathbb{E}[f^*(\mathbf{x})]$) and struggles in both the training set and any held-out test data, as it does not account for the curvature inherent in the underlying function $f^*(\mathbf{x})$ and therefore commits systematic errors.

Note that underfitting can also result from insufficient training data. Even a complex model such as a deep neural network will not perform effectively if the available training data are too limited. In this case, the model simply does not have enough information to learn the intricate patterns needed for accurate predictions, ultimately producing poor performance.

On the other hand, overfitting arises when our model is excessively complex in relation to the underlying patterns in the data. In this case, the model not only learns the true relationships present in the training set but also captures random noise, such as fitting a high-degree polynomial to data that are fundamentally linear. Because the model is capable of matching the training data so closely (to the point of perfect interpolation where $R_{\text{emp}}(\theta) = 0$), it typically exhibits low bias; however, it suffers from high variance (due to spurious noisy patterns that are only present in training set but missing in test set), causing it to perform poorly on new, unseen data. This high variance reflects the model's increased sensitivity to random fluctuations in the training set, causing it to interpret the noise as if it were a genuine signal.

To achieve a good generalization performance in the test set, we must strike an appropriate balance in the complexity of our model. This is often accomplished through regularization, which moderates overfitting by introducing an additional term that penalizes overly complex solutions. In particular, one can adopt a structural risk minimization (SRM) framework that minimizes a combined objective of the form

$$\mathcal{L}_{\text{reg}}(\theta) = \mathcal{L}_{\text{emp}}(\theta) + \lambda R(\theta),$$

where $\mathcal{L}_{\text{emp}}(\theta)$ denotes the empirical loss, $R(\theta)$ is a regularization term (e.g., the squared norm $\|\theta\|^2$ in Ridge regression), and λ is a hyperparameter controlling the strength of the penalty. By appropriately tuning λ (a problem of hyperparameter tuning) and selecting a suitable form for $R(\theta)$ (another decision outside the model estimation stage), we can manage the trade-off between fitting the training data and preventing the model from overfitting, thereby improving performance on unseen data.

2.1.2 Revisiting bias-variance tradeoff

To gain deeper insights into underfitting and overfitting, we examine the bias-variance tradeoff, a central idea in statistical learning theory that clarifies how model complexity influences the expected generalization error. Specifically, consider the expected squared prediction error for some input $\mathbf{x}$:

$$\mathbb{E}_{\mathcal{D}, \epsilon}\big[(y - \hat{f}(\mathbf{x}))^2\big],$$

where the expectation is taken over both the randomness in the training data $\mathcal{D}$ and the noise ϵ. This error can be decomposed into three components: the bias, the variance, and the irreducible error. The irreducible error term σ_ϵ^2 is due to the inherent variability of the noise ϵ and cannot be further eliminated by any model. The bias term reflects the systematic deviation between the expected model prediction and the true function value:

$$\text{Bias}[\hat{f}(\mathbf{x})] = \mathbb{E}_{\mathcal{D}}[\hat{f}(\mathbf{x})] - f^*(\mathbf{x}),$$

That is, the bias term measures how far away our prediction is from the target, on average. We would thus prefer a model with a low bias in order to fit the target as closely as possible.

Finally, the variance term captures the variability of the model predictions in different training sets and is defined as:

$$\text{Var}[\hat{f}(\mathbf{x})] \;=\; \mathbb{E}_{\mathcal{D}}\Big[\big(\hat{f}(\mathbf{x}) - \mathbb{E}_{\mathcal{D}}[\hat{f}(\mathbf{x})]\big)^{2}\Big].$$

which is essentially the variance (average squared deviation from the mean) in the model predictions. We would thus prefer to fit a stable model whose predictions are centered around the true target without too many big fluctuations.

Putting these together, it can be shown that the expected squared prediction error becomes:

$$\mathbb{E}_{\mathcal{D},\epsilon}\big[(y - \hat{f}(\mathbf{x}))^{2}\big] \;=\; \sigma_{\epsilon}^{2} \;+\; \text{Bias}^{2}[\hat{f}(\mathbf{x})] \;+\; \text{Var}[\hat{f}(\mathbf{x})].$$

Given these three terms, if the model is too simple, the bias term dominates due to a large deviation from the target on average, causing systematic errors and leading to underfitting. Conversely, if the model is overly complex, the high variance term takes over, causing it to fit random noise and leading to overfitting. Our aim is therefore to select a model of proper complexity that balances bias and variance, thereby minimizing the overall expected error.

In practice, it is often acceptable to tolerate a small increase in bias if it yields a substantial decrease in variance. A notable example is Lasso regression, where we add an l_1-norm penalty to the coefficients. While this penalty introduces bias by shrinking some coefficients toward zero, it also reduces the variance in the model's predictions through more controlled coefficient magnitudes. As a result, the overall error can be significantly lower compared to an unregularized model, demonstrating how a proper tradeoff between bias and variance can enhance a model's generalization performance.

2.2 Statistical Learning Theory

Statistical Learning Theory (SLT) is a fundamental branch of machine learning that provides a rigorous mathematical framework for understanding learning algorithms. Grounded in probability theory, statistical inference, and functional analysis, it explores questions that clarify the capabilities and limitations of predictive models, such as which learning tasks can be effectively performed in practice and which are intrinsically beyond reach. SLT also investigates assumptions about data and underlying distributions that are needed for successful learning, examining how algorithmic properties such as consistency, convergence rates, and stability affect performance. Finally, it analyzes theoretical bounds on the error rates and generalization capability of learning algorithms, thereby offering crucial insights and theoretical guarantees into how and why certain methods are effective in real-world applications. By formalizing the learning process, SLT facilitates the derivation of generalization bounds, provides deeper insights into overfitting and underfitting, and informs the development of key principles such as SRM and the bias-variance trade-off. This theoretical foundation helps us better understand how to construct models that generalize effectively while avoiding common pitfalls in practical applications.

Let us start by reviewing the basic notations. Let $\mathcal{X}$ denote the input space, which includes all possible instances or objects, and let $\mathcal{Y}$ denote the output space, which encompasses all permissible labels or responses. In binary classification, we commonly set $\mathcal{Y} = \{-1, +1\}$, so each input $\mathbf{x}_i \in \mathcal{X}$ is assigned to one of two distinct classes.

In the supervised learning setting, our goal is to determine a mapping function $f : \mathcal{X} \to \mathcal{Y}$ that can reliably map input to output. When dealing with classification, such a function f is known as a classifier. Now, given a training dataset

$$\mathcal{D}_n = \{ (\mathbf{x}_i, y_i) \}_{i=1}^n,$$

where $\mathbf{x}_i \in \mathcal{X}$ and $y_i \in \mathcal{Y}$, we seek to learn a classifier $\hat{f}$ (a predictive function) that generalizes effectively to new, unseen data. A common way to quantify this goal is to minimize the expected risk

$$R(\hat{f}) = \mathbb{E}_{(\mathbf{x},y) \sim P} \left[\mathbb{I}(y \neq \hat{f}(\mathbf{x})) \right],$$

where $\mathbb{I}(\cdot)$ denotes the indicator function, and P is the underlying data-generating process (DGP) over $\mathcal{X} \times \mathcal{Y}$ and is typically unobservable. By minimizing this expected misclassification error, we strive to ensure that $\hat{f}$ assigns the correct labels to future instances with high probability.

2.2.1 A probabilistic framework

SLT adopts a probabilistic viewpoint, assuming that the training examples $(\mathbf{x}_i, y_i)_{i=1}^n$ are independently and identically distributed (i.i.d.) samples from an unknown joint distribution P over $\mathcal{X} \times Y$.[1] Formally, we write

$$(\mathbf{x}_1, y_1), (\mathbf{x}_2, y_2), \ldots, (\mathbf{x}_n, y_n) \overset{\text{i.i.d.}}{\sim} P.$$

The i.i.d. assumption is fundamental to deriving rigorous theoretical guarantees, such as generalization bounds, because it ensures that the observed training data adequately reflect the underlying distribution.

Moreover, SLT does not impose parametric constraints on the underlying DGP P. This means that SLT operates in a distribution-free setting, making

---

[1]The theoretical analysis in SLT becomes more challenging if the data is non-i.i.d.

its results more general. Unlike conventional statistical methods that might posit a specific functional form (e.g., Gaussianity or linearity), the SLT results are valid for any DGP P. This broad applicability stems from the focus of SLT on the properties of the hypothesis space $\mathcal{F}$ (which contains all possible forms of a model in a given model class) and the learning algorithms themselves, rather than the details of how the data are generated. In particular, it deals with capacity control (e.g., VC dimension, Rademacher complexity) to analyze generalization performance. As a result, the theoretical insights gained from SLT can be applied to a wide variety of learning scenarios without making restrictive assumptions about the data-generating process.

In addition to drawing samples from the joint distribution P on $\mathcal{X} \times \mathcal{Y}$, practical labels Y often exhibit inherent randomness relative to the inputs X. That is, the mapping from an input $\mathbf{x}$ to its label y is not purely deterministic but subject to stochastic variations that can arise from two principal sources. First, label noise may creep in through errors during data collection or annotation, such as the occasional mislabeling of emails in a spam detection system. Second, class overlap may occur in situations where different classes are not clearly separable with limited feature dimensions, allowing multiple labels to be plausibly assigned to the same input. An example is attempting to infer gender solely from height, which naturally yields considerable overlap and uncertainty.

This randomness of the label is captured by the conditional probability

$$\eta(\mathbf{x}) \; = \; P\big(Y = +1 \mid X = \mathbf{x}\big),$$

which quantifies the probability that a particular input $\mathbf{x}$ belongs to the positive class. When label noise is minimal or class distinctions are sharp, $\eta(\mathbf{x})$ can be close to 0 or 1. In contrast, significant noise or pronounced overlap could drive $\eta(\mathbf{x})$ closer to 0.5, indicating greater uncertainty and posing greater challenges in training accurate classifiers. Thus, to some extent, $\eta(\mathbf{x})$ represents the entropy on the randomness of this eventual outcome (the final label class), and the learning system's goal is to generate confident probabilistic predictions whose entropy is low and the final prediction (once thresholded into a binary number) is correct.

Another key assumption in SLT is that training examples are drawn independently of the underlying distribution P. Independence implies that each sample provides unique information about P, thus enabling effective tuning and learning of the model parameters. In many real-world tasks, such as image classification with broadly varied datasets, this assumption is often reasonable. However, in certain domains, such as time series analysis or spatial data modeling, the independence assumption may not hold because of temporal or spatial correlations among data points. When such correlations exist, additional methods or modified theoretical tools are required to account for serial dependence in the data. In other words, SLT's classical guarantees are most directly applicable in scenarios where independent sampling is a valid approximation.

Besides, a further assumption in classical SLT is that the distribution P stays fixed and does not evolve over time. This stationarity condition underpins many of the standard theoretical results, ensuring that models trained on past data remain valid in the future, thus allowing us to chase a fixed target. In practice, however, real-world data distributions often shift or drift—a phenomenon known as concept drift—rendering the static assumption insufficient. As the distribution changes, a model that was once accurate can become outdated, and the model needs to either get updated frequently or incorporate an online learning element in itself in such nonstationary environments. For example, under covariate shift (changes in the distributions of predictive features), certain aspects of the data may change while others remain stable, making it challenging for models to deal with such dynamic distributions.

Lastly, the DGP P itself is presumed to be unknown during training. In practice, the learner has access only to the finite sample of data drawn from P, which serves as an empirical proxy for the true distribution. The primary aim in SLT is therefore to leverage these finite samples to make inferences about P and to build a classifier $\hat{f}$ that generalizes effectively to new, unseen data from the same source. Because this process relies on limited observations, it is inherently probabilistic (thus not guaranteed), and SLT provides a set of theoretical tools and bounds to assess the reliability of the inferred classifier in a probabilistic sense. These tools help quantify the uncertainties and potential errors in model predictions, ensuring that learning algorithms can be analyzed and compared.

2.2.2 Loss functions in supervised learning

In supervised learning, our central task is to discover a predictive function $f : \mathcal{X} \rightarrow \mathcal{Y}$ that accurately assigns the right labels y to the inputs $\mathbf{x}$.[2] To measure how well this function performs, we rely on a loss function ℓ, which specifies the cost of making a particular prediction $\hat{y}$ for a given input $\mathbf{x}$ and the corresponding target y. In binary classification, where $\mathcal{Y} = \{-1, +1\}$, one possible measure of effectiveness is the 0-1 loss:

$$\ell(\mathbf{x}, y, f(\mathbf{x})) = \begin{cases} 1 & \text{if } f(\mathbf{x}) \neq y, \\ 0 & \text{otherwise.} \end{cases}$$

This loss imposes a penalty of 1 whenever the classifier mislabels $\mathbf{x}$, thereby directly reflecting the classification accuracy. However, due to its discontinuous nature, the 0-1 loss is not amenable to standard optimization techniques (e.g., gradient-based methods). To address this, we often replace the 0-1 loss with a smooth surrogate such as the cross-entropy loss. In the binary setting $\{-1, +1\}$, a common alternative formulation is the logistic (sigmoid-based) cross-entropy:

$$\ell(\mathbf{x}, y, f(\mathbf{x})) = \log(1 + e^{-y f(\mathbf{x})}).$$

---

[2] We use a general function f instead of $\hat{f}$ as before.

Here, $f(\mathbf{x})$ can be interpreted as a real-valued predictive function (sometimes called the logit function) that maps input features in an unbounded logit, and $y \in \{-1, +1\}$. Notably, $\log\left(1 + e^{-y f(\mathbf{x})}\right)$ is differentiable with respect to the parameters of f, thereby facilitating gradient-based optimization. Moreover, from a probabilistic perspective, minimizing this cross-entropy loss corresponds to maximizing the conditional likelihood of the observed labels under a logistic model.

In regression problems, where the output space is $\mathbb{R}$, the squared error loss is a popular choice:

$$\ell(\mathbf{x}, y, f(\mathbf{x})) = \left(y - f(\mathbf{x})\right)^2,$$

which penalizes the squared difference between the predicted and true values. This encourages the model to produce more precise numerical estimates, as any deviation from the target value is penalized quadratically. This means that for model predictions with a large deviation, the resulting penalty will be even higher due to the squaring effect compared to those with a small deviation.

As with the previous section, building upon the notion of a loss function, we define the risk function $R(f)$ of a classifier f as the expected loss over the entire data-generating distribution P:

$$R(f) = \mathbb{E}_{(\mathbf{x}, y) \sim P}\left[\ell(\mathbf{x}, y, f(\mathbf{x}))\right].$$

Here, the expectation is taken with respect to all possible pairs $(\mathbf{x}, y)$ drawn from the joint distribution P. Conceptually, $R(f)$ captures how well f is expected to perform on average across the full spectrum of inputs $\mathbf{x}$ and their corresponding labels y. In practice, our goal is to choose a function f from the available training set that minimizes the empirical risk $R_{emp}(f)$, hoping that it will also perform well on unseen data (a small $R(f)$). Thus, we are essentially approximating the true risk $R(f)$ through the training samples available.

Now we introduce the Bayes classifier f_{Bayes} to represent the theoretically optimal classifier that achieves the lowest possible risk $R^* = \min_f R(f)$ for the true distribution P. In a binary classification setting where $\mathcal{Y} = \{-1, +1\}$, the Bayes classifier is defined as

$$f_{\text{Bayes}}(\mathbf{x}) := \begin{cases} +1 & \text{if } \eta(\mathbf{x}) = P(Y = +1 \mid \mathbf{x}) \geq 0.5, \\ -1 & \text{otherwise.} \end{cases}$$

Here, $\eta(\mathbf{x}) = P(Y = +1 \mid \mathbf{x})$ denotes the conditional probability of the positive class given $\mathbf{x}$.

By always predicting the class based on the highest posterior probability, the Bayes classifier minimizes the expected misclassification error, which is often referred to as the Bayes risk under the 0-1 loss. For example, if $\eta(\mathbf{x}) = 0.9$, the classifier assigns the label $+1$ because that choice has the highest probability of being correct, while if $\eta(\mathbf{x}) = 0.4$, it assigns the label -1.

Although this approach yields the theoretical optimum, it requires precise knowledge of $\eta(\mathbf{x})$ that needs to be accurately predicted using a predictive model, which is rarely available in practice. Consequently, practical algorithms strive to approximate f_{Bayes} as closely as possible using only the finite data samples at hand.

Since the true distribution P is typically unknown, we know that it is impossible to compute the Bayes classifier or the true risk $R(f)$ directly. Consequently, we resort to the empirical risk

$$R_{\text{emp}}(f) = \frac{1}{n} \sum_{i=1}^{n} \ell(\mathbf{x}_i, y_i, f(\mathbf{x}_i)),$$

which is the average loss across the finite training set. Although the empirical risk $R_{\text{emp}}(f)$ is only a proxy for the true risk, SLT provides a few methods to approximate and bound the distance between the unknown true risk $R(f)$ and its empirical estimate $R_{\text{emp}}(f)$. Drawing on tools from probability theory and statistical inference—such as concentration inequalities and uniform convergence—SLT establishes generalization bounds that relate low empirical risk to low true risk with high probability. These theoretical guarantees form the backbone of effective machine-learning algorithms, ensuring that a model that fits the training data well also achieves robust performance on novel, unseen instances, thereby bridging the gap between the theoretical Bayes classifier and the practical real-world classifier.

2.2.3 Consistency in STL

Within the framework of SLT, consistency emerges as a cornerstone property that ensures the reliability of a learning algorithm when the number of training samples becomes large. Formally, a learning algorithm is deemed consistent if the risk of its learned classifier f_n[3] converges to that of the Bayes classifier f_{Bayes}—the theoretically best possible classifier—as n approaches infinity. Mathematically, this can be expressed as

$$\lim_{n \to \infty} R(f_n) = R(f_{\text{Bayes}}),$$

where we recall that $R(f)$ denotes the expected risk of a classifier f, given by

$$R(f) = \mathbb{E}_{(\mathbf{x},y) \sim P}\big[\ell(\mathbf{x}, y, f(\mathbf{x}))\big].$$

ℓ is a loss function, and P is the underlying data-generating distribution. Consistency thus guarantees that, with sufficient data, the learned classifier f_n will asymptotically perform as well as the Bayes classifier in terms of expected loss.

---

[3] We use the subscript n to denote the fact that the prediction model f is developed based on a total of n training samples.

To analyze how an algorithm's performance evolves, we can consider a sequence of classifiers $\{f_n\}$ produced by the algorithm for increasing n. That is, we are interested in the performance of the model under consideration as the number of training samples grows. In more formal terms, we may also distinguish between weak and strong consistency based on the mode of convergence. Weak consistency requires convergence in probability, meaning for every $\epsilon > 0$,

$$\lim_{n \to \infty} P\big(|R(f_n) - R(f_{\text{Bayes}})| > \epsilon\big) = 0,$$

while strong consistency requires almost sure convergence:

$$P\Big(\lim_{n \to \infty} R(f_n) = R\big(f_{\text{Bayes}}\big) \Big) = 1.$$

Overall, whether the consistency is weak or strong, it can ensure that as the training set grows, the algorithm can effectively leverage the increased sample information, gradually reducing the gap between the learned classifier performance and the theoretically optimal Bayes risk.

Achieving consistency depends on several interrelated factors associated with the hypothesis space $\mathcal{F}$ and the learning process:

- Richness of the hypothesis space. The hypothesis space $\mathcal{F}$ must be sufficiently expressive (i.e., big enough) to encapsulate or closely approximate the Bayes classifier f_{Bayes}. Think of this relationship as constructing one point (the predictive model) to be close to another underlying true point that represents the model with Bayes risk. If $\mathcal{F}$ is overly constrained, for example, a small circle that consists of only linear functions, yet the underlying Bayes model is nonlinear and thus lives outside the circle (chosen hypothesis space), the optimal risk may be unattainable. As a result, the learned model cannot converge to the Bayes risk, thus failing to achieve consistency.

- Property of uniform convergence. To ensure that learning remains reliable for all functions in $\mathcal{F}$, the empirical risk

$$R_{\text{emp}}(f) = \frac{1}{n} \sum_{i=1}^{n} \ell\big(\mathbf{x}_i, y_i, f(\mathbf{x}_i)\big)$$

must converge uniformly to the true risk $R(f)$ as n increases. This means that $|R_{\text{emp}}(f) - R(f)|$ becomes small simultaneously for all $f \in \mathcal{F}$ with a high probability:

$$\sup_{f \in \mathcal{F}} |R_{\text{emp}}(f) - R(f)| \xrightarrow{n \to \infty} 0 \quad \text{with high probability.}$$

This requirement typically leads us to control the complexity of $\mathcal{F}$ using measures such as the Vapnik–Chervonenkis (VC) dimension (discussed later). A lower VC dimension indicates a simpler (smaller) hypothesis space and facilitates a more robust link between empirical and true risks. In contrast, if $\mathcal{F}$ is too large, there may be insufficient data to accurately estimate the performance of each hypothesis, thus hindering the convergence to the Bayes risk.

- Proper regularization. Incorporating proper regularization helps manage the balance between model complexity and fit to the training data. Mathematically, we do this by minimizing a regularized version of the empirical risk of the form:

$$\mathcal{L}_{\text{reg}}(\theta) = R_{\text{emp}}(f_\theta) + \lambda R(\theta),$$

where $R(\theta)$ is a regularization term (e.g., $\|\theta\|^2$ in Ridge regression) and λ governs the penalty strength that balances the goodness of fit versus complexity. Effective regularization prevents the learned classifier from overfitting to idiosyncratic noise in the training set, helping align its performance more closely with the Bayes optimal in the limit of large n. By suitably tuning λ and the form of $R(\theta)$, both of which are choices of hyperparameters, we can encourage the learning process to focus on meaningful structure in the data rather than irrelevant fluctuations, a key step in achieving consistency.

To sum up, consistency ensures that, as the number of training samples grows, the performance of the learning algorithm steadily improves until it effectively matches that of the optimal Bayes classifier. In practical applications, this theoretical guarantee underlies why some models often exhibit better generalization with increasing data in the training set. By reducing the risk of the learned classifier $R(f_n)$ and gradually converging to the Bayes risk $R(f_{\text{Bayes}})$, the consistency property provides a link between empirical results and theoretical optimality. However, it does not reveal how rapidly $R(f_n)$ converges to $R(f_{\text{Bayes}})$. The rate of convergence, which can be crucial when data are limited or computational resources are constrained, is a central theme in SLT's study of finite sample complexity and generalization bounds, offering more nuanced insights into the efficiency of learning algorithms.

2.2.4 Estimation–approximation trade-off

When analyzing the dynamics between empirical risk $R_{\text{emp}}(f_n)$ and true risk $R(f_n)$, it often helps to look at the estimation and approximation trade-off, which captures the balance between the expressiveness of the hypothesis space $\mathcal{F}$ and the ability of a learning algorithm to generalize from a finite training set to new, unseen instances. Balancing these factors is crucial for constructing models that not only fit the training data well but also maintain high predictive accuracy on future samples.

Formally, suppose we choose a classifier $f_n \in \mathcal{F}$ using a training set $\{(\mathbf{x}_i, y_i)\}_{i=1}^n$. Recall that our goal is to minimize the true risk $R(f_n) =$

$\mathbb{E}_{(\mathbf{x},y) \sim P}\big[\ell(\mathbf{x},y,f_n(\mathbf{x}))\big]$, where ℓ is a loss function and P is the unknown joint distribution over $\mathbf{x}$ and y. Since P is inaccessible, we approximate this objective by minimizing the empirical risk $R_{\text{emp}}(f_n) = \frac{1}{n}\sum_{i=1}^{n}\ell(\mathbf{x}_i,y_i,f_n(\mathbf{x}_i))$. When comparing the performance of f_n to that of the Bayes classifier f_{Bayes}, we can decompose this excess risk $R(f_n) - R(f_{\text{Bayes}})$[4] as follows:

$$R(f_n) - R(f_{\text{Bayes}}) = \underbrace{(R(f_n) - R(\mathcal{F}))}_{\text{Estimation Error}} + \underbrace{(R(\mathcal{F}) - R(f_{\text{Bayes}}))}_{\text{Approximation Error}},$$

where

$$R(\mathcal{F}) = \inf_{f \in \mathcal{F}} R(f)$$

is the minimal achievable risk within the prespecified hypothesis space $\mathcal{F}$. The estimation error $R(f_n) - R(\mathcal{F})$ reflects how well the learning algorithm uses the available training samples to identify the best classifier in $\mathcal{F}$, measured using the true risk. It tends to decrease as the training set grows large, provided the complexity of $\mathcal{F}$ is controlled (e.g., via regularization or bounds on the VC dimension). The approximation error $R(\mathcal{F}) - R(f_{\text{Bayes}})$ arises if $\mathcal{F}$ cannot represent or approximate the Bayes classifier accurately due to limitations in the function class, leading to irreducible bias from an overly restrictive hypothesis space. Balancing these two error components—ensuring that $\mathcal{F}$ is neither too small nor too large—lies at the heart of the estimation–approximation trade-off.

More specifically, the estimation error $R(f_n) - R(\mathcal{F})$ measures how closely the learned classifier f_n approaches the best possible classifier within the hypothesis space $\mathcal{F}$. Because we use only a finite training sample to identify f_n, the minimization of empirical risk may not perfectly recover the best model within the model class in terms of true risk. Here, the complexity of $\mathcal{F}$ plays a vital role in determining the size of this error: on the one hand, a sufficiently rich $\mathcal{F}$ can approximate the Bayes classifier more closely, potentially reducing approximation error; on the other hand, the same high capacity can inflate estimation error by allowing overfitting to the training data, making it difficult to identify a model that gives $R(\mathcal{F})$. Consequently, striking the right balance in hypothesis space design—through regularization, model selection criteria, or bounds on capacity—helps keep estimation error under control while retaining enough flexibility to capture essential patterns in the data.

The approximation error $R(\mathcal{F}) - R(f_{\text{Bayes}})$ captures how well the hypothesis space $\mathcal{F}$ is able to represent and cover the Bayes classifier f_{Bayes}. Notably, this gap arises solely from the expressive limitations of $\mathcal{F}$ and, unlike estimation error, does not diminish with increasing sample size. A more constrained hypothesis space may fail to capture intricate relationships required to mimic f_{Bayes}, thus incurring a high approximation error. Conversely, enlarging $\mathcal{F}$ can reduce approximation error but also risks elevating estimation error if the

---

[4]We put $R(f_n)$ in front since it is bigger than the true risk attained by the global minimizer f_{Bayes}.

model overfits the training data. Thus, controlling approximation error relies on selecting or designing a hypothesis space that is both sufficiently flexible to capture key patterns in the data and not so expansive as to degrade overall generalization.

The estimation–approximation trade-off thus characterizes the interplay between these two error components. Expanding the complexity of the hypothesis space $\mathcal{F}$ typically lowers approximation error by allowing the model to capture the true data-generating process more precisely. Nevertheless, an overly rich $\mathcal{F}$ can inflate estimation error because the model may overfit the limited training data, inadvertently memorizing noise rather than learning generalizable patterns. Conversely, restricting $\mathcal{F}$ can control overfitting and reduce estimation error but risks increasing approximation error if the smaller space cannot adequately cover the Bayes classifier. Balancing these concerns is a central focus in model selection and regularization, where the goal is to identify a hypothesis space that is simultaneously flexible enough to approximate the true distribution and sufficiently constrained to avoid excessive variance.

Such a trade-off is crucial for ensuring that the risk of the learned classifier converges to the Bayes risk in the limit of large sample sizes, obtaining the consistency property. If the hypothesis space $\mathcal{F}$ is excessively large—encompassing all manifestations of complex functions—it can fail to deliver consistency, meaning $R(f_n)$ may not approach $R(f_{\text{Bayes}})$ as $n \to \infty$. This issue often stems from the increased possibility of overfitting and the difficulty of uniformly approximating the true risk when $\mathcal{F}$ is extremely rich. Conversely, if $\mathcal{F}$ is too restrictive, the approximation error becomes too big, preventing the learned classifier from ever covering the Bayes classifier. Hence, selecting a hypothesis space with the right balance of expressive power and regularization is fundamental to securing both the flexibility needed to capture critical structures in the data and the consistency required for generalization.

Figure 2.1 illustrates the tradeoff between approximation and estimation error. In particular, when we minimize the empirical risk over a finite training set, we obtain a classifier f_{emp}. Although f_{emp} can match the training data closely, it may be prone to overfitting, hence diverging from the ideal solution measured using the true risk. Meanwhile, the best-in-class model f_F is constrained by the expressive capacity of the chosen hypothesis space F, determining how effectively it can approximate the Bayes classifier f_{Bayes}. In this case, no matter how we optimize over the training data, the intrinsic limitations of F fix the gap between f_F and f_{Bayes}. Consequently, there is a balance to be struck: while enlarging F can reduce approximation error, doing so risks inflating estimation error if f_{emp} becomes overly tuned to the idiosyncrasies (coupled with noise) of the training set.

This estimation–approximation tradeoff also aligns closely with the classical bias-variance tradeoff in the previous section. Roughly speaking, the approximation error in the estimation–approximation perspective corresponds to bias aspect in the bias-variance paradigm: high bias arises when the

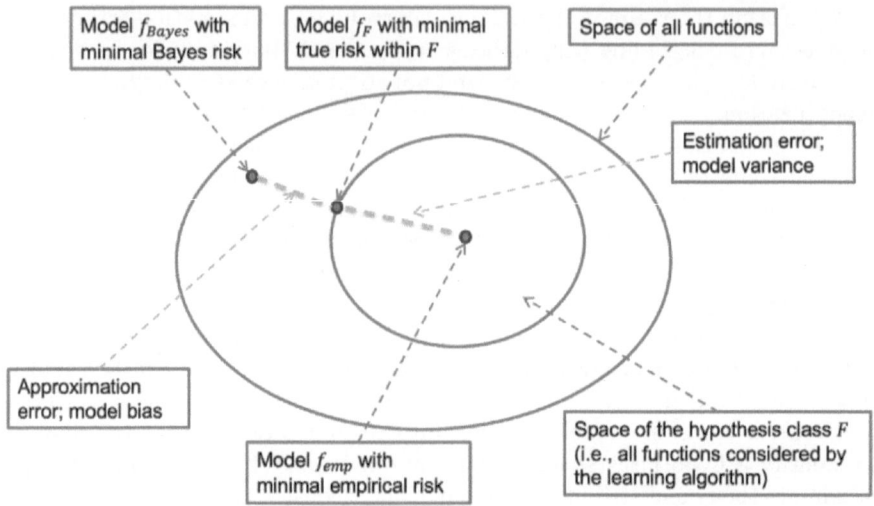

FIGURE 2.1
Illustrating the approximation and estimation error.

hypothesis space is too limited to fully capture the underlying data structure, resulting in underfitting. Meanwhile, the estimation error parallels variance: high variance implies that the model is overly sensitive to fluctuations in the training set, thus overfitting and failing to generalize. Recognizing this duality is helpful for guiding both the theoretical analysis and practical design of learning algorithms, as it underscores that any move to reduce bias (or approximation error) by enlarging the hypothesis space must be balanced against the potential for increased variance (or estimation error), unless appropriate regularization is applied.

As illustrated in Figure 2.2, the interplay between estimation error and approximation error becomes evident when varying the size (or complexity) of the hypothesis class. If the hypothesis space is constrained to be small, the chosen model class may be too limited to learn the intrinsic patterns in the data, resulting in underfitting. In this regime, approximation error dominates because the model cannot capture essential structures, while the estimation error can remain relatively small due to lower variance in fitting a simpler function class. Conversely, if the hypothesis class is extremely rich and big, the model can nearly match even intricate data relationships, thus reducing approximation error. However, this expressive power also heightens the likelihood of overfitting, causing estimation error to grow as the model becomes highly sensitive to idiosyncrasies in the training data. Striking a balance between these two extremes—managing the flexibility of the model while preventing undue variance—lies at the core of developing robust, generalizable learning systems.

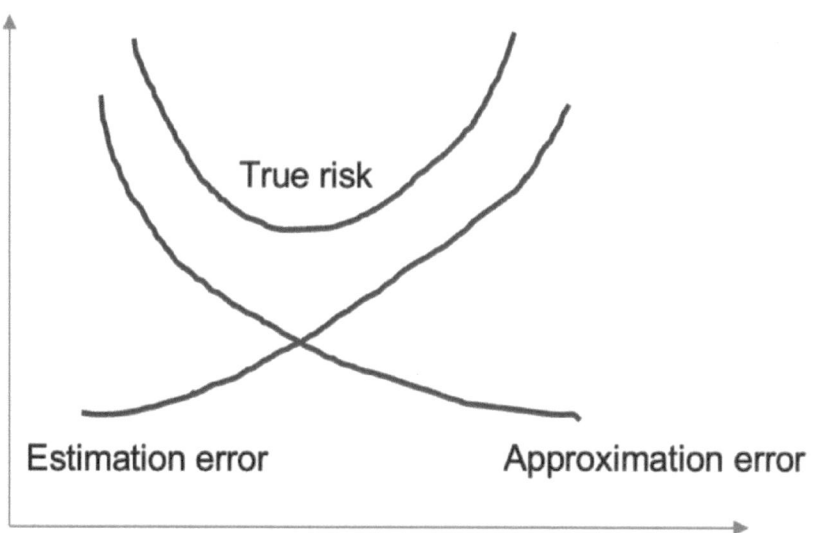

FIGURE 2.2
Illustrating the tradeoff between approximation and estimation error.

Thus, achieving a low total error (in terms of the true risk) necessitates a balance in which the hypothesis space is sufficiently expressive to reduce bias (or approximation error), yet not so expansive that the model becomes vulnerable to large variance (or estimation error). Strategies such as regularization and model selection criteria—notably SRM—are key tools in SLT that help navigate this trade-off effectively. In particular, selecting an appropriately sized and structured hypothesis space $\mathcal{F}$ is an important decision. An overly large $\mathcal{F}$ can contain models complex enough to perfectly fit the training data yet fail to generalize, driving up estimation error due to overfitting. Conversely, if $\mathcal{F}$ is too restricted, it cannot capture the true patterns in the data, leading to pronounced approximation error. By carefully calibrating the flexibility of $\mathcal{F}$, one seeks to strike the sweet spot where both errors are sufficiently controlled, thereby achieving robust generalization.

To help navigate the estimation–approximation trade-off, regularization methods restrict the complexity of learned classifiers, thus controlling the estimation error without excessively increasing the approximation error. For example, we can vary the complexity of the hypothesis space $\mathcal{F}$ and select the model that demonstrates the most promising generalization performance on the training data. By decomposing the risk into estimation and approximation components, SLT offers a clear framework for analyzing how the choices of hypothesis space and learning strategies influence the ultimate performance of the classifiers. Properly managing this trade-off is critical for minimizing overall risk and securing consistency, wherein the classifier's expected risk converges to that of the Bayes optimal solution given sufficiently large sample sizes. Consequently, SLT not only illuminates why particular algorithms generalize well or poorly but also guides the design of methods that balance complexity and parsimony to achieve reliable, data-driven learning.

2.2.5   Empirical risk minimization

In the basic supervised learning setup, we have a training dataset $\{(\mathbf{x}_i, y_i)\}_{i=1}^n$, where each input $\mathbf{x}_i \in \mathcal{X}$ is paired with an output $y_i \in \mathcal{Y}$. Our task is to learn a classifier $f : \mathcal{X} \to \mathcal{Y}$ that minimizes the true risk:

$$R(f) = \mathbb{E}\big[\ell(\mathbf{x}, y, f(\mathbf{x}))\big],$$

where ℓ is a loss function that measures how far the prediction $f(\mathbf{x})$ deviates from the true label y. Since the underlying distribution $P(X, Y)$ is unknown, we cannot compute $R(f)$ directly.

Instead, the principle of Empirical Risk Minimization (ERM) tackles this difficulty by substituting the true risk $R(f)$ with the empirical risk $R_{\mathrm{emp}}(f)$ as the proxy,[5] which is the average loss computed over the given training data:

$$R_{\mathrm{emp}}(f) = \frac{1}{n} \sum_{i=1}^n \ell\big(\mathbf{x}_i, y_i, f(\mathbf{x}_i)\big).$$

In accordance with ERM, we then define our learned classifier f_n as the function (the best model measured in terms of empirical risk) in the hypothesis space $\mathcal{F}$ that minimizes this empirical risk:

$$f_n := \arg \min_{f \in \mathcal{F}} R_{\mathrm{emp}}(f).$$

Here, $\mathcal{F}$ represents the set of candidate classifiers we allow ourselves to consider, which can range from simpler, more constrained families to highly flexible function classes. By choosing f_n to minimize $R_{\mathrm{emp}}(f)$, we aim to approximate the best possible classifier under P, even though we can only access a finite sample from that distribution.

The principle behind ERM finds a key theoretical underpinning in the Law of Large Numbers (LLN), a cornerstone of probability theory. The LLN states that, given a set of i.i.d. random variables $\{\xi_i\}_{i=1}^n$, their sample average converges to the expected value (the first moment of $P(\xi)$) as n grows large:

$$\frac{1}{n} \sum_{i=1}^n \xi_i \longrightarrow \mathbb{E}[\xi] \quad \text{as } n \to \infty.$$

---

[5]In practice, we further divide the available data into training, validation, and testing, with the latter two serving as the proxy for true risk.

In the ERM framework, we treat the empirical risk $R_{\text{emp}}(f)$ as the sample average of the losses $\ell(\mathbf{x}_i, y_i, f(\mathbf{x}_i))$ for all instances $i \in \{1, \ldots, n\}$. For any fixed classifier f, the LLN ensures that

$$R_{\text{emp}}(f) = \frac{1}{n} \sum_{i=1}^{n} \ell(\mathbf{x}_i, y_i, f(\mathbf{x}_i)) \longrightarrow \mathbb{E}\big[\ell(\mathbf{x}, y, f(\mathbf{x})\big] = R(f) \quad \text{as } n \to \infty.$$

Thus, for sufficiently large n, the empirical risk $R_{\text{emp}}(f)$ becomes a good approximation of the true risk $R(f)$ for that particular f. Indeed, when exposed to all possible variations of training data, the model should be able to interpolate perfectly and recover the true underlying model with Bayes risk. This convergence property ensures that minimizing $R_{\text{emp}}(f)$ can effectively approximate minimizing $R(f)$, provided that the hypothesis f is not overly complex and the sample size is substantial.

To evaluate the effectiveness of using the empirical risk $R_{\text{emp}}(f)$ as an estimator of the true risk $R(f)$ for a fixed classifier f, we can invoke classical concentration results like Hoeffding's inequality. Specifically, suppose the loss values $\ell(\mathbf{x}_i, y_i, f(\mathbf{x}_i))$ are bounded in the interval $[0, 1]$. Then, Hoeffding's inequality states that for any $\epsilon > 0$,

$$P\Big(\big|R_{\text{emp}}(f) - R(f)\big| \geq \epsilon\Big) \leq 2\exp\big(-2\,n\,\epsilon^2\big),$$

implying that the likelihood of $R_{\text{emp}}(f)$ deviating from $R(f)$ by more than ϵ vanishes exponentially fast with increasing sample size n. Concretely, for sufficiently large n, the empirical risk $R_{\text{emp}}(f)$ offers a good approximation of the true risk $R(f)$ with high probability. This exponential convergence rate is a key theoretical reason why optimizing the empirical risk can yield reliable results in practice, provided the classifier remains fixed during the analysis and the training set is large enough.

A significant complication emerges when we try to apply Hoeffding's inequality to ERM: the selected classifier f_n depends on the training data itself. In contrast to a fixed, data-independent classifier, f_n is specifically optimized to minimize the empirical risk on the observed training samples, thereby violating the independence assumption crucial to concentration inequalities like Hoeffding's. Consequently, although such inequalities assure us that the empirical and true risks align closely for any single, fixed classifier, they do not automatically carry over to the data-dependent classifier f_n.

This gap can also manifest itself in overfitting, where the empirical risk $R_{\text{emp}}(f_n)$ could be deceptively small, but the true risk $R(f_n)$ is substantially larger. Thus, while ERM strives to choose a classifier that appears optimal based on the training set, the data dependence of f_n introduces subtleties that require more refined theoretical tools. Subsequent chapters address these issues in detail, examining how regularization, SRM, and other advanced techniques help ensure that the ERM framework remains consistent and avoids pitfalls such as overfitting, ultimately supporting robust and generalizable machine-learning models.

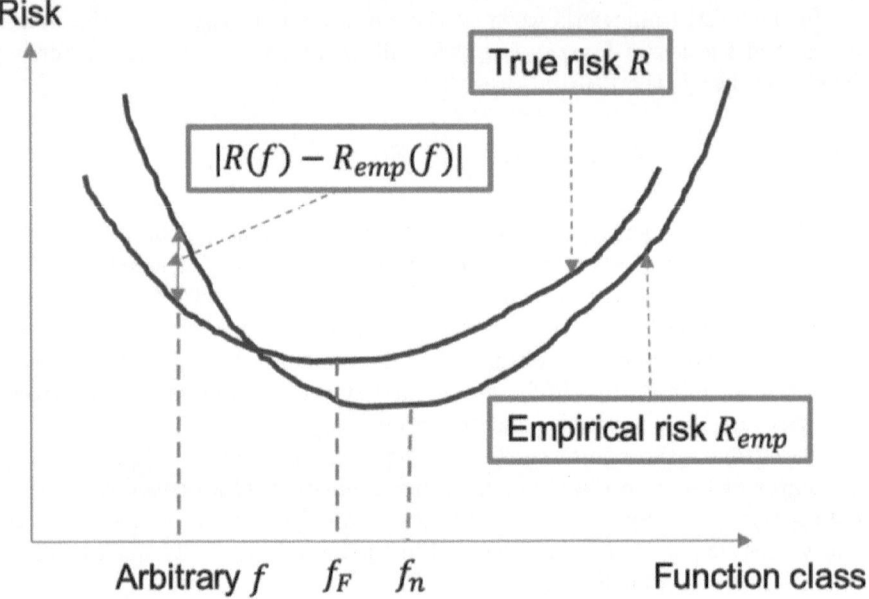

Risk

$|R(f) - R_{emp}(f)|$

True risk R

Empirical risk R_{emp}

Arbitrary f f_F f_n Function class

FIGURE 2.3
Convergence of empirical risk to actual risk.

2.2.6   Uniform convergence

To ensure that ERM is consistent, we must place additional constraints on the hypothesis space $\mathcal{F}$. The fundamental insight from the VC theory is that ERM's consistency hinges on the worst-case performance[6] across all classifiers $f \in \mathcal{F}$ that the learning algorithm might pick. This leads to a more rigorous variant of the LLN, the Uniform Law of Large Numbers (ULLN), which ensures that the empirical risk converges uniformly to the true risk over the entire hypothesis space.

In Figure 2.3, we see a schematic showing how the empirical risk converges to the true risk as the number of training samples increases. The x-axis represents a simplified, one-dimensional projection of $\mathcal{F}$. Each point along this axis corresponds to a distinct function, which might be a random function f, the empirical risk minimizer f_n, or the best classifier f_F within $\mathcal{F}$. If the global Bayes classifier f_{Bayes} belongs to $\mathcal{F}$, then $f_F = f_{\text{Bayes}}$. Otherwise, $\mathcal{F}$ merely contains an approximation to the optimal solution. By enforcing conditions such as the finite VC dimension or other capacity constraints, we can apply ULLN to show that, as the training sample size grows, the empirical risks of all functions in $\mathcal{F}$ converge to their respective true risks. Consequently, the empirical minimizer f_n will, with high probability, be close to the best

---

[6]We note that this principle plays a similar role as in robust optimization.

possible classifier f_F in $\mathcal{F}$. This uniform convergence property underlies the consistency of ERM, enabling it to yield classifiers whose performance steadily approaches that of the best realizable classifier in the hypothesis class as more data become available.

As depicted, we have two curves representing the true risk R and the empirical risk R_{emp}. The classifier f_n that minimizes R_{emp} is generally distinct from f_F, the minimizer of the true risk within the hypothesis space $\mathcal{F}$. For any specific classifier $f \in \mathcal{F}$, we use the gap $|R(f) - R_{\text{emp}}(f)|$ to reflect how well the empirical risk approximates the true risk for f.

Now, the standard LLN guarantees that, for a fixed f, $R_{\text{emp}}(f)$ converges to $R(f)$ as n grows. However, this does not ensure that every function in $\mathcal{F}$ simultaneously achieves the same level of convergence. To achieve uniform convergence across all hypotheses in $\mathcal{F}$, we can resort to ULLN, which states that for any $\epsilon > 0$, there exists a sufficiently large n such that

$$P\left(\sup_{f \in \mathcal{F}} |R(f) - R_{\text{emp}}(f)| \geq \epsilon\right) \longrightarrow 0 \quad \text{as } n \to \infty.$$

This property says that, with high probability, all functions in $\mathcal{F}$ exhibit a negligible gap between their empirical and true risks once n is large enough. Such property is critical to establish that empirical minimization leads to a classifier f_n whose risk closely approximates the true minimizer f_F in $\mathcal{F}$, thus facilitating the consistency of empirical minimization.

To elaborate on the details of establishing the consistency of ERM using ULLN, let us consider the difference between the true risk of the classifier selected by ERM, f_n, and the true risk of the best classifier within the hypothesis space $\mathcal{F}$, denoted $f_{\mathcal{F}}$. This difference is given by

$$R(f_n) - R(f_{\mathcal{F}}).$$

Because $f_{\mathcal{F}}$ minimizes the true risk over $\mathcal{F}$, we have $R(f_n) \geq R(f_{\mathcal{F}})$, ensuring that the difference is nonnegative. To analyze this excess risk further, we add and subtract the corresponding empirical risks, yielding

$$\begin{aligned}
R(f_n) - R(f_{\mathcal{F}}) = &\left[R(f_n) - R_{\text{emp}}(f_n)\right] \\
&+ \left[R_{\text{emp}}(f_n) - R_{\text{emp}}(f_{\mathcal{F}})\right] \\
&+ \left[R_{\text{emp}}(f_{\mathcal{F}}) - R(f_{\mathcal{F}})\right].
\end{aligned}$$

Since f_n is chosen to minimize the empirical risk over $\mathcal{F}$, it follows that

$$R_{\text{emp}}(f_n) \leq R_{\text{emp}}(f_{\mathcal{F}}),$$

which implies that the middle term satisfies

$$R_{\text{emp}}(f_n) - R_{\text{emp}}(f_{\mathcal{F}}) \leq 0.$$

Thus, we can bound the excess risk by

$$R(f_n) - R(f_{\mathcal{F}}) \leq \left[R(f_n) - R_{\text{emp}}(f_n)\right] + \left[R_{\text{emp}}(f_{\mathcal{F}}) - R(f_{\mathcal{F}})\right].$$

By the ULLN, for any $\epsilon > 0$, there exists a sufficiently large sample size n such that, with high probability,

$$\sup_{f \in \mathcal{F}} |R(f) - R_{\mathrm{emp}}(f)| \leq \epsilon.$$

This uniform bound immediately implies that

$$R(f_n) - R_{\mathrm{emp}}(f_n) \leq \epsilon \quad \text{and} \quad R_{\mathrm{emp}}(f_\mathcal{F}) - R(f_\mathcal{F}) \leq \epsilon.$$

Substituting these inequalities into our bound on the excess risk, we obtain

$$R(f_n) - R(f_\mathcal{F}) \leq \epsilon + \epsilon = 2\epsilon.$$

Now, let us introduce a threshold $\delta > 0$ and set $\delta = 2\epsilon$. Then, the above bound implies that if the uniform convergence holds at the level $\epsilon = \delta/2$, we have

$$R(f_n) - R(f_\mathcal{F}) < \delta.$$

The key statement we now incorporate is as follows: if, on the other hand, we observe that

$$R(f_n) - R(f_\mathcal{F}) \geq \delta,$$

then necessarily

$$\sup_{f \in \mathcal{F}} |R(f) - R_{\mathrm{emp}}(f)| \geq \frac{\delta}{2}.$$

In words, a large excess risk (at least δ) can occur only if the uniform deviation between true and empirical risks over $\mathcal{F}$ is at least $\delta/2$.

Since event $\{R(f_n) - R(f_\mathcal{F}) \geq \delta\}$ is a subset of the event $\{\sup_{f \in \mathcal{F}} |R(f) - R_{\mathrm{emp}}(f)| \geq \frac{\delta}{2}\}$, we know that the probability of the former cannot exceed the probability of the latter, thus giving:

$$P\left(R(f_n) - R(f_\mathcal{F}) \geq \delta\right) \leq P\left(\sup_{f \in \mathcal{F}} |R(f) - R_{\mathrm{emp}}(f)| \geq \frac{\delta}{2}\right).$$

Since the left-hand side is a nonnegative quantity by definition, we can also write

$$P\left(|R(f_n) - R(f_\mathcal{F})| \geq \delta\right) \leq P\left(\sup_{f \in \mathcal{F}} |R(f) - R_{\mathrm{emp}}(f)| \geq \frac{\delta}{2}\right).$$

As the sample size n increases, the ULLN ensures that the probability on the right-hand side becomes arbitrarily small. Hence, the probability that the excess risk exceeds δ also becomes negligible. Therefore, the deterministic bound on the excess risk, together with the uniform convergence guarantee provided by the ULLN, leads directly to the probabilistic bound, which demonstrates that the excess risk of the ERM-selected classifier diminishes with increasing sample size, thereby establishing the consistency of ERM. In other words, the true risk of the classifier selected by ERM converges to the risk of the best possible classifier within $\mathcal{F}$. This result establishes the consistency of ERM, as it guarantees that, with enough data, the performance of the ERM-selected classifier will approach that of the optimal classifier in the hypothesis space.

2.3 Capacity Measures of Functional Class

So far, we have examined the properties of the function space that determine the consistency of the ERM principle, particularly focusing on whether it will eventually work "in the limit" as the number of training samples grows to infinity. As discussed earlier, this long-term behavior is captured by the concept of uniform convergence, which guarantees that the empirical risk converges uniformly to the true risk across the entire hypothesis space. However, since we invariably operate with a finite set of data points in practice, it becomes crucial to look at how uniform convergence manifests in finite-sample scenarios. This deeper exploration not only enables us to derive precise bounds on the excess risk incurred by the learned classifier but also offers valuable insights into the intrinsic characteristics of function classes, such as their complexity measures like the VC dimension, that promote uniform convergence. Understanding these properties is essential for designing and selecting hypothesis spaces that effectively balance complexity and generalization, ensuring that the empirical performance reliably reflects the true risk even with a limited amount of data.

To better analyze how well the empirical risk approximates the true risk uniformly over the entire hypothesis space $\mathcal{F}$, we begin by considering the worst-case probability

$$P\left(\sup_{f \in \mathcal{F}} |R(f) - R_{\text{emp}}(f)| > \epsilon\right).$$

This expression measures the likelihood that there exists at least one function f in $\mathcal{F}$ for which the difference between the true risk $R(f)$ and the empirical risk $R_{\text{emp}}(f)$ exceeds a small threshold ϵ. In other words, it quantifies the worst-case deviation across all classifiers, a key component in understanding uniform convergence.

To effectively bound this probability, we can employ two essential analytical tools. First, we can use the union bound, which allows us to break down the probability over the entire (possibly infinite) collection of functions into a sum of probabilities for individual events. This decomposition makes the overall bound more tractable. Second, we leverage the method of symmetrization by introducing a ghost sample, a separate set of data drawn independently from the same distribution. This technique transforms the problem by replacing the complex dependency structure of the original sample with a symmetrized version, thereby facilitating tighter and more manageable probabilistic estimates.

Together, these methods enable us to derive meaningful finite-sample guarantees on the uniform convergence of the empirical risk to the true risk, which are crucial for understanding the generalization properties of learning algorithms operating under the ERM framework.

Let us start with the union bound.

2.3.1   Union bound for finite function classes

The union bound is a straightforward yet powerful tool that enables us to extend the standard LLN from individual functions to a ULLN over a finite set of functions. To illustrate, consider a finite hypothesis space $\mathcal{F} = \{f_1, f_2, \ldots, f_m\}$. For each function $f_i \in \mathcal{F}$, we can apply the standard LLN in the form of the Chernoff bound, which guarantees that

$$P\Big(|R(f_i) - R_{\mathrm{emp}}(f_i)| \geq \epsilon\Big) \leq 2\exp(-2n\epsilon^2).$$

Our objective is to extend these individual concentration bounds to obtain a uniform bound over the entire set $\mathcal{F}$. To do this, note that the event

$$\left\{\sup_{f \in \mathcal{F}} |R(f) - R_{\mathrm{emp}}(f)| \geq \epsilon\right\}$$

occurs if and only if there exists at least one function f_i in $\mathcal{F}$ for which $|R(f_i) - R_{\mathrm{emp}}(f_i)| \geq \epsilon$. By the union bound, we have:

$$P\Big(\sup_{f \in \mathcal{F}} |R(f) - R_{\mathrm{emp}}(f)| \geq \epsilon\Big) = P\Big(\bigcup_{i=1}^{m} \{|R(f_i) - R_{\mathrm{emp}}(f_i)| \geq \epsilon\}\Big)$$

$$\leq \sum_{i=1}^{m} P\Big(|R(f_i) - R_{\mathrm{emp}}(f_i)| \geq \epsilon\Big).$$

Substituting the Chernoff bound for each term in the sum yields

$$\sum_{i=1}^{m} P\Big(|R(f_i) - R_{\mathrm{emp}}(f_i)| \geq \epsilon\Big) \leq 2m\exp(-2n\epsilon^2).$$

This inequality effectively transforms individual concentration bounds into a uniform concentration bound in the finite function class $\mathcal{F}$. The factor m here represents the size, or capacity, of the hypothesis space; if $\mathcal{F}$ is fixed and finite, m is a constant. Consequently, as the sample size n increases, the term $2m\exp(-2n\epsilon^2)$ decays exponentially to zero. This rapid decay ensures that the empirical risk $R_{\mathrm{emp}}(f)$ uniformly converges to the true risk $R(f)$ for all functions in $\mathcal{F}$.

In summary, by applying the union bound together with the Chernoff bound, we obtain a uniform concentration inequality that guarantees, with high probability, the maximum deviation between the empirical and true risks over $\mathcal{F}$ becomes arbitrarily small as n grows. This uniform convergence is fundamental to establishing the consistency of ERM over a finite function class. In particular, it implies that the ERM-selected classifier f_n converges to the best classifier $f_{\mathcal{F}}$ in terms of true risk as $n \to \infty$, ensuring that the learning algorithm performs reliably even when only a finite number of training samples is available.

2.3.2 Extending beyond finite function classes

While the union bound offers an elegant and effective solution for finite function classes, many real-world learning problems involve infinite or exceedingly large hypothesis spaces. In such cases, the straightforward application of the union bound would lead to an exponential dependence on the size of the function class m, which is impractical when m is infinite. Obviously, simply using m fails to present a meaningful bound when it tends toward infinity. To circumvent this issue, we need additional analytical techniques to replace the combinatorial factor m with more refined and tractable measures of capacity. One influential approach, introduced by Vapnik and Chervonenkis, leverages the concept of a ghost sample to extend these uniform convergence results to infinite function classes. This method allows us to substitute the raw count m with alternative capacity measures, such as the VC dimension or the covering numbers, that effectively quantify the "size" or "complexity" of the function class to a finite number, even when the function class itself is unbounded.

A central technique in this framework is known as symmetrization, which proceeds by incorporating a ghost sample. Suppose that we are given a set of training samples $\{(\mathbf{x}_i, y_i)\}_{i=1}^n$, where each $(\mathbf{x}_i, y_i)$ is drawn independently of the underlying distribution P. To facilitate symmetrization, we introduce an i.i.d. ghost sample set $\{(\mathbf{x}_i', y_i')\}_{i=1}^n$, drawn from the same distribution P and independent of the original sample. This ghost sample is a theoretical construct used solely for analysis; in practice, there is no need to generate it physically. By comparing the empirical risks computed on these two independent samples, symmetrization allows us to derive uniform convergence bounds that hold over the entire (potentially infinite) function class $\mathcal{F}$.

The introduction of ghost samples and the application of symmetrization enable us to move beyond the limitations of the union bound when dealing with infinite function classes. By replacing the combinatorial dependence on m with more nuanced capacity measures, these techniques provide guarantees of uniform convergence, which, in turn, underpin the consistency and generalization performance of learning algorithms operating over rich hypothesis spaces.

Using the ghost sample, we define the empirical risk on this additional sample set as

$$R'_{\text{emp}}(f) = \frac{1}{n} \sum_{i=1}^{n} \ell(\mathbf{x}'_i, y'_i, f(\mathbf{x}'_i)).$$

This construct is valuable because it allows us to replace the unobservable true risk $R(f)$ with a comparison between two empirical risks, $R_{\text{emp}}(f)$ and $R'_{\text{emp}}(f)$, both of which can be calculated directly from the data. We can then think of a way to connect the excess risk between $R(f)$ and $R_{\text{emp}}(f)$ with the new excess risk between $R_{\text{emp}}(f)$ and $R'_{\text{emp}}(f)$. Since both the original sample $\{(\mathbf{x}_i, y_i)\}_{i=1}^{n}$ and the ghost sample $\{(\mathbf{x}'_i, y'_i)\}_{i=1}^{n}$ are drawn independently from the same distribution P, they serve as two independent estimators of the true risk. This is crucial because ULLN ensures that each of these empirical risks converges to $R(f)$ as the sample size increases.

The power of the ghost sample is fully realized in the following Symmetrization Lemma. For any $\epsilon > 0$, the lemma states that

$$P\left(\sup_{f \in \mathcal{F}} \left|R(f) - R_{\text{emp}}(f)\right| > \epsilon\right) \leq 2\,P\left(\sup_{f \in \mathcal{F}} \left|R_{\text{emp}}(f) - R'_{\text{emp}}(f)\right| > \frac{\epsilon}{2}\right).$$

In this inequality, the probability on the left-hand side is taken with respect to the distribution of the original sample, while the probability on the right-hand side considers the joint distribution of both the original sample and the ghost sample. Essentially, the lemma tells us that the worst-case deviation between the true risk and the empirical risk over the function class $\mathcal{F}$ can be bounded by twice the probability that the difference between two independently computed empirical risks exceeds $\epsilon/2$.

This result stands because the ghost sample, by virtue of being an independent draw from the same distribution P, mirrors the behavior of the true risk in expectation. Consequently, any substantial deviation between the empirical risk and the true risk must be reflected in the discrepancy between the two independent empirical estimates. Thus, by controlling the latter, we obtain a uniform bound on the former. This symmetrization technique transforms the challenge of bounding the unobservable true risk into a problem of analyzing the difference between two computable quantities using the available training set, thus providing a powerful tool for establishing uniform convergence results even when the function class $\mathcal{F}$ is infinite.

The symmetrization lemma becomes intuitive once we observe that both $R_{\text{emp}}(f)$ and $R'_{\text{emp}}(f)$ serve as unbiased estimators of the true risk $R(f)$. Since the original and ghost samples are independently drawn, the difference in the empirical risk, expressed as

$$R_{\text{emp}}(f) - R'_{\text{emp}}(f)$$

captures the inherent variability in the empirical estimation process. Intuitively, if the empirical risks computed from two independent samples are close for every function $f \in \mathcal{F}$, it strongly suggests that both estimates are also close to the unobservable true risk $R(f)$. This observation is the cornerstone that allows us to bound the excess risk $|R(f) - R_{\mathrm{emp}}(f)|$.

While a full proof is beyond our current scope, the essence of the symmetrization lemma can be grasped through standard symmetrization techniques in probability theory, notably the use of Rademacher variables. To outline the approach, we can introduce a sequence of independent Rademacher random variables $\{\sigma_i\}_{i=1}^n$, each taking the value $+1$ or -1 with probability 0.5. With these variables, the difference between the two empirical risks can be re-expressed as

$$R_{\mathrm{emp}}(f) - R'_{\mathrm{emp}}(f) = \frac{1}{n} \sum_{i=1}^n \sigma_i \left[\ell(\mathbf{x}_i, y_i, f(\mathbf{x}_i)) - \ell(\mathbf{x}'_i, y'_i, f(\mathbf{x}'_i)) \right].$$

This re-expressed formulation makes sense since multiplying each difference by σ_i essentially randomizes the signs of these differences and does not change the distribution due to symmetry. The properties of Rademacher variables, combined with concentration inequalities, enable us to relate the probability of large deviations in the original empirical process to that of the symmetrized process. In effect, this transformation reduces the problem of bounding

$$\sup_{f \in \mathcal{F}} |R(f) - R_{\mathrm{emp}}(f)|$$

which depends on the unobservable true risk $R(f)$, to a problem involving only observable empirical quantities computed on the original and ghost samples.

Furthermore, the lemma implies that when analyzing an infinite hypothesis class $\mathcal{F}$, one can effectively limit the analysis to a finite set of distinct function behaviors on a combined sample of size $2n$. In particular, since each function in $\mathcal{F}$ assigns a label to each of the $2n$ instances, there are at most 2^{2n} distinct patterns that the functions can produce. This finite combinatorial bound on the number of distinct behaviors allows us to replace the raw capacity size m in the union bound with more sophisticated capacity measures such as the VC dimension or covering numbers, which remain applicable even when $\mathcal{F}$ is infinite. Ultimately, by relating the uniform deviation between the true and empirical risks to the deviation between two independent empirical estimates, the symmetrization lemma provides a powerful pathway to establish uniform convergence and, hence, the consistency of learning algorithms under ERM.

2.3.3 Shattering coefficient

To quantify the capacity of the function class $\mathcal{F}$ and derive meaningful generalization bounds, we can rely on a combinatorial measure known as the shattering coefficient. For a given sample $Z = \{\mathbf{x}_1, \mathbf{x}_2, \ldots, \mathbf{x}_n\}$, let $\mathcal{F}_Z$ denote the set of distinct mappings of the functions in $\mathcal{F}$ to the sample Z; that is, two functions $f, g \in \mathcal{F}$ are considered distinct on Z if there exists at least one point $\mathbf{x}_i \in Z$ such that $f(\mathbf{x}_i) \neq g(\mathbf{x}_i)$. The shattering coefficient $\mathcal{N}(\mathcal{F}, n)$ is then defined as the maximum number of distinct functions (or equivalently, distinct labels) that $\mathcal{F}$ can achieve over any sample of size n:

$$\mathcal{N}(\mathcal{F}, n) = \max_{Z \subset \mathcal{X}, |Z| = n} |\mathcal{F}_Z|.$$

In other words, the shattering coefficient $\mathcal{N}(\mathcal{F}, n)$ measures the maximum number of distinct dichotomies (i.e., ways to assign labels) that the function class $\mathcal{F}$ can produce on any set of n points. This definition captures the richness or complexity of $\mathcal{F}$; for example, if $\mathcal{N}(\mathcal{F}, n) = 2^n$, it indicates that $\mathcal{F}$ is capable of identifying all possible 2^n binary labels (as 1 or 0) in some sample of size n, meaning that $\mathcal{F}$ shatters that sample. A higher shattering coefficient thus suggests a more complex function class that can represent a larger variety of patterns, though at the potential cost of overfitting, while a lower shattering coefficient indicates a simpler hypothesis space with possibly better generalization but a reduced capacity to fit complex data. Consequently, the shattering coefficient provides a critical tool for bounding the growth of the function class with respect to the sample size and plays a central role in establishing uniform convergence results and generalization bounds in learning theory.

The shattering coefficient and the VC dimension provide a combinatorial measure of complexity that comes in handy when bounding the generalization error. In many uniform convergence results, the growth of the shattering coefficient with respect to n appears in the upper bounds for the difference between the empirical risk and the true risk. For example, if the shattering coefficient grows polynomially with n, then one can often derive tighter bounds on the uniform deviation than if it were exponential. This insight helps to explain why models with excessively high complexity (i.e., a very large shattering coefficient) may overfit the training data, leading to poor generalization performance.

Returning to our goal of bounding the probability

$$P \left(\sup_{f \in \mathcal{F}} |R(f) - R_{\mathrm{emp}}(f)| > \epsilon \right),$$

We note that even though the function class $\mathcal{F}$ may be infinite, when considering a combined sample of size $2n$ only a finite subset of distinct functions, denoted by $\mathcal{F}_{Z_{2n}}$, is relevant. Here, $\mathcal{F}_{Z_{2n}}$ represents the set of distinct classifications that functions in $\mathcal{F}$ can exhibit on the combined sample Z_{2n}. This observation allows us to simplify the analysis and obtain the bound

$$P \left(\sup_{f \in \mathcal{F}} |R(f) - R_{\mathrm{emp}}(f)| > \epsilon \right) \leq 2P \left(\sup_{f \in \mathcal{F}_{Z_{2n}}} |R_{\mathrm{emp}}(f) - R'_{\mathrm{emp}}(f)| > \frac{\epsilon}{2} \right),$$

where $R'_{\mathrm{emp}}(f)$ is the empirical risk calculated on an independent ghost sample of size n.

For each fixed function $f \in \mathcal{F}_{Z_{2n}}$, assuming that the loss function ℓ is bounded within $[0, 1]$, the difference $R_{\text{emp}}(f) - R'_{\text{emp}}(f)$ consists of independent random variables bounded in the interval $[-1, 1]$. Applying Hoeffding's inequality to each such function, we have

$$P\left(|R_{\text{emp}}(f) - R'_{\text{emp}}(f)| > \frac{\epsilon}{2}\right) \leq 2 \exp\left(-\frac{2\left(\frac{\epsilon}{2}\right)^2 n}{(1 - (-1))^2}\right) = 2 \exp\left(-\frac{n\epsilon^2}{8}\right).$$

Next, by applying the union bound over all functions in $\mathcal{F}_{Z_{2n}}$, we obtain

$$P\left(\sup_{f \in \mathcal{F}_{Z_{2n}}} |R_{\text{emp}}(f) - R'_{\text{emp}}(f)| > \frac{\epsilon}{2}\right) \leq 2\mathcal{N}(\mathcal{F}, 2n) \exp\left(-\frac{n\epsilon^2}{8}\right),$$

where $\mathcal{N}(\mathcal{F}, 2n)$ is the shattering coefficient, representing the maximum number of distinct functions that $\mathcal{F}$ can produce on any sample of size $2n$.

Integrating this bound with the symmetrization lemma, we arrive at

$$P\left(\sup_{f \in \mathcal{F}} |R(f) - R_{\text{emp}}(f)| > \epsilon\right) \leq 2 \times 2\mathcal{N}(\mathcal{F}, 2n) \exp\left(-\frac{n\epsilon^2}{8}\right)$$

$$= 4\mathcal{N}(\mathcal{F}, 2n) \exp\left(-\frac{n\epsilon^2}{8}\right).$$

This final bound indicates that, as the sample size n increases, the probability that the uniform deviation between the true risk and the empirical risk exceeds ϵ decays exponentially, provided that the shattering coefficient $\mathcal{N}(\mathcal{F}, 2n)$ does not grow too rapidly (e.g., if it grows polynomially with n). This exponential decay is fundamental to establish the consistency of ERM, ensuring that the empirical risk minimizer f_n converges to the best-in-class risk minimizer $f_{\mathcal{F}}$ in $\mathcal{F}$ as $n \to \infty$.

To illustrate this with an example, consider the class of linear classifiers in $\mathbb{R}^d$:

$$\mathcal{F} = \{f_{\mathbf{w}}(\mathbf{x}) = \text{sign}(\mathbf{w}^\top \mathbf{x}) \mid \mathbf{w} \in \mathbb{R}^d\}.$$

For this function class, the shattering coefficient can be bounded by

$$\mathcal{N}(\mathcal{F}, n) \leq \sum_{k=0}^{d} \binom{n}{k} \leq n^d.$$

This polynomial growth in n contrasts sharply with the exponential growth 2^n that would occur if all possible labels were realizable. Consequently, the probability bound for this example becomes

$$P\left(\sup_{f \in \mathcal{F}} |R(f) - R_{\text{emp}}(f)| > \epsilon\right) \leq 4n^{2d} \exp\left(-\frac{n\epsilon^2}{8}\right),$$

which still decays exponentially with n for any fixed dimension d. This bound underscores how controlling the capacity of the function class through the shattering coefficient is essential to ensure that ERM generalizes well from finite samples to the underlying distribution.

2.3.4   Different function classes

Now we examine how different growth rates of the shattering coefficient $\mathcal{N}(\mathcal{F}, n)$ influence the consistency of ERM by considering various function classes characterized by distinct shattering behaviors and discussing their implications for uniform convergence and generalization.

For example, consider function classes where the shattering coefficient grows sub-polynomially with the sample size n. Examples of such classes include certain regularized linear models or simple decision trees with bounded depth, for which we might have

$$\mathcal{N}(\mathcal{F}, n) \leq \log^k(n) \quad \text{for some constant } k.$$

Substituting this bound into our previously derived inequality

$$P\left(\sup_{f \in \mathcal{F}} |R(f) - R_{\text{emp}}(f)| > \epsilon\right) \leq 4\mathcal{N}(\mathcal{F}, 2n) \exp\left(-\frac{n\epsilon^2}{8}\right),$$

we obtain

$$P\left(\sup_{f \in \mathcal{F}} |R(f) - R_{\text{emp}}(f)| > \epsilon\right) \leq 4 \log^k(2n) \exp\left(-\frac{n\epsilon^2}{8}\right).$$

Because $\log^k(2n)$ grows very slowly compared to the exponential decay $\exp\left(-\frac{n\epsilon^2}{8}\right)$, as n increases the exponential term dominates. This ensures that the entire expression tends to zero rapidly. In this case, not only is ERM consistent, but the convergence is very fast: the probability that the supremum deviation exceeds ϵ vanishes rapidly as the sample size grows, guaranteeing a reliable generalization.

Next, consider function classes with a linear growth of the shattering coefficient, such as certain kernel-based methods with fixed kernel parameters. For these classes, one can often establish a bound of the form

$$\mathcal{N}(\mathcal{F}, n) \leq cn \quad \text{for some constant } c.$$

Substituting this into the same inequality yields

$$P\left(\sup_{f \in \mathcal{F}} |R(f) - R_{\text{emp}}(f)| > \epsilon\right) \leq 4\mathcal{N}(\mathcal{F}, 2n) \exp\left(-\frac{n\epsilon^2}{8}\right) \leq 4c(2n) \exp\left(-\frac{n\epsilon^2}{8}\right),$$

which simplifies to

$$P\left(\sup_{f \in \mathcal{F}} |R(f) - R_{\text{emp}}(f)| > \epsilon\right) \leq 8cn \exp\left(-\frac{n\epsilon^2}{8}\right).$$

Although the term $8cn$ grows linearly with n, the exponential decay $\exp\left(-\frac{n\epsilon^2}{8}\right)$ still dominates for large n. Thus, as the sample size increases, the overall bound tends to zero, albeit at a slower rate compared to the sub-polynomial case. Nonetheless, ERM remains consistent for function classes with linear shattering coefficients, with the probability of the supremum deviation exceeding ϵ decreasing exponentially fast.

In both cases, the growth rate of the shattering coefficient is pivotal for determining the rate at which uniform convergence occurs. For function classes with sub-polynomial growth, the exponential decay in the probability bound is even more pronounced, leading to rapid convergence and strong generalization. For classes with linear growth, while the convergence is still exponential, the rate is somewhat moderated by the linear factor. However, the key takeaway here is that as long as the shattering coefficient does not grow too quickly (e.g., if it grows most polynomially with n), the overall probability that the supremum deviation exceeds any fixed ϵ will decay exponentially with the sample size. This exponential decay is crucial for ensuring that the empirical risk minimizer f_n converges to the true risk minimizer $f_{\mathcal{F}}$ within the function class, thereby establishing the consistency of ERM in a rigorous, quantitative manner.

Advanced models, such as neural networks with bounded architectures or ensemble methods like random forests with a fixed number of trees, may exhibit shattering coefficients that grow super-polynomially but sub-exponentially with the sample size. For instance, for such function classes we might have

$$\mathcal{N}(\mathcal{F}, n) \leq \exp(n^\alpha) \quad \text{for some } 0 < \alpha < 1.$$

Substituting this bound into our uniform convergence inequality yields

$$4\mathcal{N}(\mathcal{F}, 2n) \exp\left(-\frac{n\epsilon^2}{8}\right) \leq 4 \exp\left((2n)^\alpha\right) \exp\left(-\frac{n\epsilon^2}{8}\right) = 4 \exp\left((2n)^\alpha - \frac{n\epsilon^2}{8}\right).$$

Here, the term $\exp\left((2n)^\alpha\right)$ grows faster than any polynomial in n but slower than a full exponential, since $\alpha < 1$. As n increases, the linear term $-\frac{n\epsilon^2}{8}$ in the exponent eventually dominates $(2n)^\alpha$. This means that despite the relatively rapid growth of $\exp\left((2n)^\alpha\right)$, the overall expression is ultimately controlled by the stronger exponential decay of $\exp\left(-\frac{n\epsilon^2}{8}\right)$. Consequently, for function classes with super-polynomial but sub-exponential shattering coefficients, the probability bound still tends to zero as $n \to \infty$, albeit at a slower convergence rate than for classes with polynomial or logarithmic shattering coefficients. This result implies that the ERM remains consistent in these cases: the empirical risk minimizer f_n will converge to the best classifier in $\mathcal{F}$ as more data are collected.

In contrast, consider the function class $\mathcal{F}_{\text{all}}$, which encompasses all possible classifiers. For this class, the shattering coefficient is maximal:

$$\mathcal{N}(\mathcal{F}_{\text{all}}, n) = 2^n.$$

Substituting this into the uniform convergence bound gives

$$4\mathcal{N}(\mathcal{F}_{\text{all}}, 2n) \exp\left(-\frac{n\epsilon^2}{8}\right) = 4\,2^{2n} \exp\left(-\frac{n\epsilon^2}{8}\right) = 4 \exp\left(2n \ln 2 - \frac{n\epsilon^2}{8}\right).$$

The term $4\,2^{2n}$ grows exponentially with n, and when we simplify the exponent, we obtain

$$2n \ln 2 - \frac{n\epsilon^2}{8} = n\left(2\ln 2 - \frac{\epsilon^2}{8}\right).$$

For the entire bound to tend to zero, we require the coefficient $2\ln 2 - \frac{\epsilon^2}{8}$ to be negative; that is,

$$2\ln 2 - \frac{\epsilon^2}{8} < 0 \quad \Rightarrow \quad \epsilon^2 > 16\ln 2 \approx 11.09.$$

However, in typical learning scenarios, ϵ is chosen to be much smaller than $\sqrt{16\ln 2} \approx 3.33$. As a result, the exponent $2n\ln 2 - \frac{n\epsilon^2}{8}$ remains positive, leading the probability bound to increase exponentially with n. This indicates that for function classes with exponential shattering coefficients, ERM is not consistent; the empirical risk minimizer may not generalize effectively, as the probability that the uniform deviation between the true and empirical risks exceeds ϵ fails to vanish.

In summary, the growth rate of the shattering coefficient $\mathcal{N}(\mathcal{F}, n)$ is an important component in determining the consistency of ERM. For function classes with sub-exponential (or polynomial) growth, the exponential decay in the probability bound guarantees that the uniform deviation shrinks to zero with increasing sample size, ensuring that ERM generalizes well. Conversely, for function classes with exponential growth in their shattering coefficient, the bound does not decay, and may even grow, indicating a failure of consistency. This analysis underscores the importance of controlling the capacity of the function class (measured using shattering coefficient $\mathcal{N}(\mathcal{F}, n)$) to achieve reliable generalization in machine learning.

Beyond the categories discussed earlier, many practical function classes exhibit intermediate growth rates in their shattering coefficients, often shaped by additional constraints or regularization strategies. For example, regularized models—which incorporate L1 or L2 penalties—can effectively reduce the shattering coefficient by limiting the capacity of the hypothesis space. Such regularization essentially enforces simpler models that are less prone to overfitting, ensuring that $\mathcal{N}(\mathcal{F}, n)$ grows at most polynomially or even subpolynomially with n.

Similarly, sparse models, such as sparse linear models or those employing feature selection methods, tend to have reduced shattering coefficients. By constraining the number of active parameters or selected features, these models effectively control their complexity, leading to more favorable generalization properties. In this way, sparsity serves as a natural mechanism to restrict the effective size of the function class, thereby promoting consistent performance as more data becomes available.

Hierarchical models, such as deep neural networks, present another interesting case. Their shattering coefficients can vary significantly depending on architectural choices such as depth, width, and the implementation of regularization techniques. By carefully designing these networks, for example, by bounding the depth or limiting connectivity, one can achieve manageable growth rates of the shattering coefficient. This controlled complexity is essential to ensure that ERM remains consistent, even for complex, multilayer architectures.

In general, in real-world applications, the balance between expressiveness and complexity is delicate. By selecting or designing models whose shattering coefficients grow at most polynomially with the sample size, we can harness the power of ERM (in-sample) while ensuring robust generalization performance (out-of-sample). Techniques such as regularization, sparsity enforcement, and careful architectural design are useful strategies for maintaining this balance, ultimately leading to more reliable and consistent learning outcomes.

2.3.5 Generalization bounds

In the analysis of ERM, it is often useful to reinterpret the existing probability bounds to better understand the relationship between empirical risk $R_{\text{emp}}(f)$ and true risk $R(f)$. That is, rather than fixing a deviation threshold ϵ and computing the probability that the difference $|R(f) - R_{\text{emp}}(f)|$ exceeds ϵ, we can invert the perspective by fixing a confidence level $\delta > 0$ and then determining how small ϵ must be to ensure that the true risk is close to the empirical risk with high probability.

Specifically, starting with the probability bound

$$P\left(\sup_{f \in \mathcal{F}} |R(f) - R_{\text{emp}}(f)| > \epsilon\right) \leq 2\mathcal{N}(\mathcal{F}, 2n) \exp\left(-\frac{n\epsilon^2}{4}\right),$$

our goal is to express this bound in terms of ϵ given a fixed probability level δ. To achieve this, we set the right-hand side equal to δ:

$$2\mathcal{N}(\mathcal{F}, 2n) \exp\left(-\frac{n\epsilon^2}{4}\right) = \delta.$$

Solving for ϵ involves taking logarithms on both sides. Specifically, we have:

$$\exp\left(-\frac{n\epsilon^2}{4}\right) = \frac{\delta}{2\mathcal{N}(\mathcal{F}, 2n)},$$

which, upon taking the natural logarithm, gives

$$-\frac{n\epsilon^2}{4} = \ln\left(\frac{\delta}{2\mathcal{N}(\mathcal{F}, 2n)}\right).$$

Rearranging and solving for ϵ yields

$$\epsilon = \sqrt{\frac{4}{n} \ln \left(\frac{2\mathcal{N}(\mathcal{F}, 2n)}{\delta} \right)}.$$

Substituting this expression for ϵ back into the probability statement, we obtain the generalization bound:

$$P\left(\sup_{f \in \mathcal{F}} |R(f) - R_{\mathrm{emp}}(f)| > \epsilon \right) \leq \delta, \quad \text{where} \quad \epsilon = \sqrt{\frac{4}{n} \left(\ln(2\mathcal{N}(\mathcal{F}, 2n)) - \ln(\delta) \right)}.$$

This bound says that, with probability at least $1 - \delta$, every function $f \in \mathcal{F}$ satisfies

$$R(f) \leq R_{\mathrm{emp}}(f) + \sqrt{\frac{4}{n} \left(\ln(2\mathcal{N}(\mathcal{F}, 2n)) - \ln(\delta) \right)}.$$

This generalization bound is instrumental in establishing the consistency of ERM. In particular, ERM is consistent for the function class $\mathcal{F}$ if the term

$$\sqrt{\frac{\ln(2\mathcal{N}(\mathcal{F}, 2n))}{n}}$$

converges to zero as the sample size n tends to infinity. This condition is typically satisfied when the shattering coefficient $\mathcal{N}(\mathcal{F}, 2n)$ grows at most polynomially with n. In such cases, the logarithmic growth in the numerator is eventually outweighed by the linear growth in the denominator, which ensures that ϵ decreases as n increases.

Moreover, the bound holds uniformly for all functions $f \in \mathcal{F}$, ensuring that the empirical risk minimizer f_n (which minimizes $R_{\mathrm{emp}}(f)$) also satisfies this inequality. Notably, the bound is robust in the sense that it applies to any function in the hypothesis class $\mathcal{F}$, regardless of how it is selected. However, since the bound must hold uniformly over $\mathcal{F}$, it is a conservative statement, especially for large or complex function classes. In such cases, the bound on ϵ may not be tight (meaning there is still room for improvement), and the effectiveness of the bound heavily depends on the growth rate of $\mathcal{N}(\mathcal{F}, 2n)$. For function classes with rapidly increasing shattering coefficients, the bound may become vacuous, which means consistency cannot be achieved even if n goes to infinity.

In summary, by reinterpreting the probability bound in terms of a fixed confidence level δ and solving for the deviation ϵ, we obtain a concrete generalization bound that quantifies how close the true risk is to the empirical risk with high probability. This bound is central to understanding the conditions under which the ERM is consistent and emphasizes the importance of controlling the complexity of the function class to ensure robust generalization as the sample size grows.

2.3.6 VC dimension

In the preceding sections, we formulated generalization bounds in terms of the shattering coefficient $\mathcal{N}(\mathcal{F}, n)$. Although these bounds are powerful, directly evaluating the shattering coefficient can be challenging because of its combinatorial nature. To overcome this difficulty, researchers have developed various alternative capacity measures, each with its own set of advantages and limitations. Among these measures, the VC dimension stands out as one of the most prominent and widely used. The VC dimension provides a succinct numerical summary that characterizes the growth behavior of the shattering coefficient, thereby facilitating the analysis of learning algorithms in the context of ERM.

The VC dimension of a function class $\mathcal{F}$, denoted by $\text{VC}(\mathcal{F})$, is a fundamental concept in SLT. It quantifies the capacity of $\mathcal{F}$ by capturing the largest number of points that the function class can shatter. Recall that a sample $Z_n = \{z_1, z_2, \ldots, z_n\}$ is said to be shattered by $\mathcal{F}$ if, for every possible binary labeling of the sample, there exists a function in $\mathcal{F}$ that correctly classifies the sample according to that labeling. Formally, Z_n is shattered by $\mathcal{F}$ if

$$|\mathcal{F}_{Z_n}| = 2^n,$$

where $\mathcal{F}_{Z_n}$ denotes the set of all distinct binary pointwise labelings induced by functions in $\mathcal{F}$ on the sample Z_n. The VC dimension is then defined as the largest integer d for which there exists at least one sample of size d that is shattered by $\mathcal{F}$. If no finite d exists, the VC dimension is defined to be infinite:

$$\text{VC}(\mathcal{F}) = \max\left\{ n \in \mathbb{N} \,\middle|\, |\mathcal{F}_{Z_n}| = 2^n \text{ for some } Z_n \subset \mathcal{X} \right\}.$$

For example, in the case of linear classifiers in $\mathbb{R}^d$, the VC dimension is $d+1$. This implies that there exists a set of $d+1$ points that can be shattered by linear classifiers, but no set of $d+2$ points can be shattered. Similarly, decision trees with a bounded depth k have a finite VC dimension that depends on both k and the number of features available.

A cornerstone result connecting the VC dimension to the shattering coefficient is provided by the Sauer-Shelah Lemma. Independently discovered by Sauer (1972), Shelah (1972), and Vapnik and Chervonenkis (1971), this lemma offers a tight combinatorial bound on $\mathcal{N}(\mathcal{F}, n)$ based on the VC dimension d.[7] Specifically, it states that if $\mathcal{F}$ has a finite VC dimension d, then for all $n \in \mathbb{N}$:

$$\mathcal{N}(\mathcal{F}, n) \leq \sum_{i=0}^{d} \binom{n}{i}.$$

Moreover, for $n \geq d$, this bound can be further simplified to

$$\mathcal{N}(\mathcal{F}, n) \leq \left(\frac{en}{d}\right)^d,$$

---

[7]We also used this result in the proof in the previous section.

where e is the base of the natural logarithm. This result is important because it shows that if the VC dimension is finite, the shattering coefficient grows at most polynomially with n. As a consequence, by substituting this bound into our earlier generalization bounds, we deduce that the term

$$\sqrt{\frac{\ln(2\,\mathcal{N}(\mathcal{F}, 2n))}{n}}$$

tends to zero as n increases, thereby ensuring that the ERM is consistent for function classes with finite VC dimension.

On the other hand, if $\mathcal{F}$ has an infinite VC dimension, then it can shatter arbitrarily large samples, and the shattering coefficient grows exponentially with n (i.e., $\mathcal{N}(\mathcal{F}, n) = 2^n$). In such cases, the generalization bounds derived from the shattering coefficient do not guarantee convergence, implying that ERM may not generalize well.

Thus the VC dimension serves as a crucial tool for translating the combinatorial complexity captured by the shattering coefficient into a more manageable numerical form. This, in turn, provides a pathway to establish generalization bounds and guarantees the consistency of ERM for function classes whose capacity is appropriately controlled. Understanding these relationships is fundamental for designing and selecting models that balance expressiveness and complexity, thereby ensuring robust learning outcomes in practice.

Combining the insights from the Sauer-Shelah Lemma and the derived generalization bounds, we arrive at a fundamental theorem that links the VC dimension of a function class $\mathcal{F}$ to the consistency of ERM. Specifically, the theorem states that ERM is consistent with respect to a function class $\mathcal{F}$ if and only if the VC dimension $\text{VC}(\mathcal{F})$ is finite. In other words, if the function class has a finite VC dimension, then the empirical risk minimizer converges to the true risk minimizer as the sample size grows, ensuring reliable generalization.

The proof of this theorem can be divided into two cases. First, when $\text{VC}(\mathcal{F})$ is finite, the Sauer-Shelah Lemma guarantees that the shattering coefficient $\mathcal{N}(\mathcal{F}, n)$ grows at most polynomially with n. Substituting this polynomial growth into our generalization bounds shows that the deviation ϵ between the true risk $R(f)$ and the empirical risk $R_{\text{emp}}(f)$ decreases as n increases. Consequently, the uniform convergence of the empirical risk to the true risk is ensured, and ERM converges to the optimal risk minimizer within $\mathcal{F}$. In the second case, when $\text{VC}(\mathcal{F})$ is infinite, the shattering coefficient $\mathcal{N}(\mathcal{F}, n)$ grows exponentially with n. This exponential growth renders the generalization bound ineffective because the deviation ϵ does not tend to zero, implying that the empirical risk minimizer may not converge to the true risk minimizer, and ERM fails to generalize consistently.

An important aspect of both the VC dimension and the shattering coefficient is their distribution independence, in that they depend solely on the structure of the function class $\mathcal{F}$ and are agnostic to the underlying data distribution P. This universality is advantageous because it ensures that the theoretical generalization bounds hold uniformly across all possible data distributions, making the results robust and widely applicable. However, this same distribution independence can be a double-edged sword. By not taking into account the specific properties of the data distribution, the VC dimension can sometimes yield overly conservative bounds. In practice, this can lead to loose generalization bounds, especially in scenarios where the data distribution is particularly benign or structured in a way that allows for tighter bounds. Moreover, for certain function classes, the VC dimension might overestimate the true complexity needed to effectively model the data, resulting in conservative estimates that do not accurately reflect the model's practical generalization performance.

In summary, the theorem connecting the VC dimension to the consistency of ERM provides a clear criterion: ERM is consistent if and only if the function class has finite capacity as measured by the VC dimension. This result not only underpins much of the theoretical framework of statistical learning but also guides practical model selection and regularization strategies aimed at balancing expressiveness and complexity to achieve reliable generalization in real-world applications.

2.3.7 Rademacher complexity

Although the shattering coefficient $\mathcal{N}(\mathcal{F}, n)$ and the VC dimension $\mathrm{VC}(\mathcal{F})$ are foundational measures for quantifying the capacity of a function class, they sometimes fall short in generating tight generalization bounds. To address these challenges, researchers have introduced another useful capacity measure known as the Rademacher complexity, which is briefly mentioned in the previous section. Unlike the shattering coefficient and VC dimension, which are purely combinatorial and distribution-independent, the Rademacher complexity now takes into account the underlying probability distribution of the data, thereby allowing it to yield sharper and more refined generalization bounds that better reflect the actual behavior of learning algorithms in practice.

The Rademacher complexity quantifies the ability of a function class $\mathcal{F}$ to fit random noise. It is defined using Rademacher variables, $\sigma_1, \sigma_2, \ldots, \sigma_n$, which are independent random variables that take the values $+1$ and -1 with equal probability 0.5. These variables can be thought of as the unique outcomes of fair coin tosses and provide a mechanism to introduce randomness into the analysis. Formally, given a sample set $Z_n = \{X_1, X_2, \ldots, X_n\}$, the Rademacher complexity of $\mathcal{F}$ is defined as

$$\mathcal{R}(\mathcal{F}) = \mathbb{E}_\sigma \left[\sup_{f \in \mathcal{F}} \frac{1}{n} \sum_{i=1}^n \sigma_i f(X_i) \right],$$

where $\mathbb{E}_\sigma$ denotes the expectation over the Rademacher variables. Intuitively, this definition measures, on average, how well the functions in $\mathcal{F}$ fit random labeling in the data. In other words, if there exists a function in $\mathcal{F}$ that can strongly correlate with random signs σ_i, then the Rademacher complexity will be high, signaling that the function class has a high capacity to fit noise, which is a potential red flag for overfitting.

To gain further intuition, consider fixing a particular realization of the Rademacher variables. Each σ_i can be interpreted as a randomly assigned label to the corresponding data point X_i. The product $\sigma_i f(X_i)$ is positive if f aligns with the random label at X_i and negative otherwise. The sum $\sum_{i=1}^{n} \sigma_i f(X_i)$ aggregates the classification accuracy over the entire sample. Taking the supremum over all functions in $\mathcal{F}$ identifies the function that best fits these random labels, and the expectation over σ averages this maximal alignment over all possible random labelings. In this way, a high value of $\mathcal{R}(\mathcal{F})$ indicates that the function class can easily fit random noise, which in turn reflects a high capacity and a potential risk of overfitting. Conversely, a lower Rademacher complexity implies that the class is more constrained, which can potentially lead to better generalization performance.

The generalization bound that utilizes Rademacher complexity is typically expressed as

$$P\left(R(f) \le R_{\text{emp}}(f) + 2\,\mathcal{R}(\mathcal{F}) + \sqrt{\frac{\log(1/\delta)}{2n}}\right) \ge 1 - \delta.$$

This inequality asserts that, with probability at least $1 - \delta$, the true risk $R(f)$ of any function $f \in \mathcal{F}$ is bounded above by its empirical risk $R_{\text{emp}}(f)$ plus an additional term that scales with twice the Rademacher complexity and a confidence term $\sqrt{\frac{\log(1/\delta)}{2n}}$. The term $2\,\mathcal{R}(\mathcal{F})$ captures the complexity of the function class in a data-dependent manner, thereby offering a more customized measure and often a tighter bound than the VC dimension. Moreover, this bound holds uniformly over all functions in $\mathcal{F}$, making it a strong guarantee for the performance of the empirical risk minimizer.

Rademacher complexity offers several distinct advantages. First, because it is distribution-dependent, it can adapt to the specific characteristics of the data, often yielding tighter bounds than those based solely on the VC dimension or shattering coefficient. Second, its formulation is flexible enough to be applied across a wide variety of function classes, ranging from linear models to deep neural networks, where traditional measures such as the VC dimension may be difficult to compute or interpret. Finally, by being amenable to empirical estimation, the Rademacher complexity provides practical insights into the capacity of a model, guiding model selection and regularization design. Techniques such as L1 (Lasso) and L2 (ridge) regularization, for example, inherently aim to reduce the Rademacher complexity by constraining the size or sparsity of the model parameters, thereby reducing the risk of overfitting and promoting better generalization. In neural network architectures, choices regarding depth, width, and connectivity all influence the Rademacher complexity, and careful design can help maintain a manageable capacity even in complex models.

In summary, Rademacher complexity is a powerful, distribution-sensitive capacity measure that enhances our ability to derive sharp generalization bounds. It is distribution dependent since we need to draw samples $Z_n = \{X_1, X_2, \ldots, X_n\}$ from the underlying distribution P. Thus the resulting Rademacher complexity depends on the specific characteristics of the data distribution. This means that if the distribution P is such that the data are concentrated in regions where the functions in $\mathcal{F}$ do not vary much, the Rademacher complexity may be lower than if the data were more spread out across the input space. Thus, by quantifying the capacity of a function class to fit random noise, it provides a direct link between model complexity and generalization performance, making it an invaluable tool in both the theoretical analysis and practical application of machine-learning algorithms.

2.3.8 Other generalization bounds and capacity concepts

While we have already explored foundational capacity measures, including the shattering coefficient $\mathcal{N}(\mathcal{F}, n)$, the VC dimension $\mathrm{VC}(\mathcal{F})$, and the Rademacher complexity $\mathcal{R}(\mathcal{F})$, the literature encompasses a wide array of other capacity concepts that provide additional measures of function class complexity. Although detailing each of these additional measures is beyond the scope of this chapter, it is instructive to note that most generalization bounds in SLT share a common structural form. This unified form can be expressed as:

$$P\Big(R(f) \leq R_{\mathrm{emp}}(f) + \mathrm{capacity}(\mathcal{F}) + \mathrm{confidence}(\delta)\Big) \geq 1 - \delta.$$

In this formulation, the probability statement asserts that with confidence at least $1 - \delta$, the true risk $R(f)$ of any function $f \in \mathcal{F}$ is bounded above by its empirical risk $R_{\mathrm{emp}}(f)$ plus two additional terms: a capacity term and a confidence term. The empirical risk term $R_{\mathrm{emp}}(f)$ quantifies how well the function f fits the observed training data, serving as a direct measure of performance on the given sample. The capacity term $\mathrm{capacity}(\mathcal{F})$ indicates the complexity or richness of the function class $\mathcal{F}$. It can be expressed using different capacity measures such as the VC dimension, Rademacher complexity, or others, each of which captures distinct aspects of the function class's potential to fit data. A higher capacity indicates a more expressive (or complex) function class, which might lead to better training performance but also increases the risk of overfitting. Finally, the confidence term $\mathrm{confidence}(\delta)$ incorporates the chosen probability level δ and typically depends on both δ and the sample size n. This term ensures that the bound holds with high probability over different samples drawn from the underlying data distribution.

A critical aspect of these generalization bounds is their worst-case nature. Because they are designed to hold uniformly for all functions in $\mathcal{F}$, they necessarily account for the "most badly behaved" function within the class. This worst-case analysis corresponds to a robust optimization objective, although it also tends to yield conservative (and sometimes loose) bounds that may not tightly reflect the performance of the functions typically selected by learning algorithms like ERM.

To illustrate the interplay of these components, consider how different capacity measures influence the bounds. When using the VC dimension $\mathrm{VC}(\mathcal{F})$ as the capacity measure, the generalization bound may take the form (with probability $\geq 1 - \delta$):

$$R(f) \leq R_{\mathrm{emp}}(f) + \sqrt{\frac{\mathrm{VC}(\mathcal{F}) \cdot \log\left(\frac{n}{\mathrm{VC}(\mathcal{F})}\right)}{n}} + \sqrt{\frac{\log(1/\delta)}{2n}}.$$

In this bound, the capacity term $\sqrt{\frac{\mathrm{VC}(\mathcal{F}) \cdot \log\left(\frac{n}{\mathrm{VC}(\mathcal{F})}\right)}{n}}$ reflects the complexity of the function class in relation to the sample size. Alternatively, when employing the Rademacher complexity $\mathcal{R}(\mathcal{F})$, which is sensitive to the data distribution and often yields sharper bounds, the generalization bound becomes:

$$R(f) \leq R_{\mathrm{emp}}(f) + 2\,\mathcal{R}(\mathcal{F}) + \sqrt{\frac{\log(1/\delta)}{2n}}.$$

This formulation benefits from the distribution-dependent nature of the Rademacher complexity, frequently resulting in tighter, less conservative bounds compared to those derived solely from the VC dimension. In general, the general form of these bounds accommodates various capacity measures, allowing us to tailor the analysis to the specific properties of the function class and the data distribution at hand.

Also, note that these uniform guarantees apply across the entire function class and thus present strong theoretical foundations. However, it is important to recognize that, in practice, the classifiers selected by methods like ERM are rarely the worst-case functions considered in these bounds. As a result, while these theoretical guarantees are robust and broadly applicable, they may be conservative relative to the actual performance observed in real-world applications. Nonetheless, understanding these generalization bounds is fundamental, as they offer deep insights into the trade-offs between model capacity, sample size, and confidence, and they underpin the rigorous analysis of learning algorithms in SLT.

2.4   Summary

In this chapter, we introduced the foundational concepts of SLT, providing the basic mathematical framework for understanding the finite-sample performance of learning algorithms under the ERM framework. We began by introducing key concepts such as the true risk, defined as the expected loss over the unknown data-generating distribution, and the empirical risk, which serves as a proxy computed over a finite training dataset.

A central theme of this chapter is the formulation and analysis of generalization bounds, which quantify the gap between the empirical risk and the true risk. We introduced several capacity measures, including the shattering coefficient, VC dimension, and Rademacher complexity, all of which are designed to capture the expressive power (or capacity) of a function class (or hypothesis class). These measures play a crucial role in determining whether the ERM principle can guarantee consistency; that is, whether the model selected by minimizing the empirical risk converges to the optimal (Bayes) classifier as the sample size increases. In particular, the Sauer-Shelah Lemma provided a combinatorial bridge between the VC dimension and the growth rate of the shattering coefficient, revealing that if the VC dimension is finite, the shattering coefficient grows at most polynomially with the sample size, which in turn ensures that the generalization error diminishes as more data become available.

We also explored the estimation–approximation trade-off, which highlights the delicate balance between the complexity of the hypothesis space and the ability to generalize from finite samples. A function class that is too simple incurs high approximation error (or bias), while an overly complex class is prone to high estimation error (or variance) due to overfitting. This trade-off is analogous to the classical bias-variance trade-off and is central to designing learning algorithms that strike an effective balance between fitting the training data and maintaining robust predictive performance on unseen data. Regularization techniques, such as L1 and L2 penalties, were discussed as practical strategies for controlling this balance by reducing the effective capacity of the model.

Furthermore, the chapter highlighted the role of uniform convergence and the ULLN in establishing the consistency of ERM. By ensuring that the empirical risk converges uniformly to the true risk over the entire hypothesis space, uniform convergence guarantees that the empirical risk minimizer not only performs well on the training data but also generalizes reliably to the true data distribution. This convergence was analyzed using various probabilistic tools, including concentration inequalities like Hoeffding's inequality, and further refined by methods such as symmetrization, which facilitate the extension of these results to infinite function classes.

Overall, the insights presented in this chapter lay the theoretical groundwork for understanding the interplay between model complexity, sample size, and generalization performance. By linking capacity measures to generalization bounds, SLT provides robust theoretical guarantees that guide the design and evaluation of learning algorithms. This foundation is essential for advancing both the theory and practice of machine learning, ensuring that models developed in finite-sample settings can perform effectively in real-world applications.

3

Classical Perspectives on Generalization

In this chapter, we will examine the comprehensive training process of one of the most fundamental models in supervised learning: linear regression. In this framework, linear regression constructs a linear mapping from input features to an output target, thereby serving as a critical foundation for more complex modeling endeavors in our exploration toward model generalization. Through an in-depth analysis of a simple linear regression model's training process, we aim to obtain a better understanding of the essential components of training, including the data, model architecture, objective (or cost) function, and the optimization process.

Our discussion will delve into the underlying mathematical intuition associated with each component, which is indispensable for a nuanced understanding of various modeling operations. Moreover, we will contextualize these theoretical concepts by underscoring their practical relevance, emphasizing that effective learning is closely tied to a clear comprehension of the context in which these models operate. To bridge the gap between theory and practice, we will also incorporate implementation details to allow for better appreciation of both conceptual and practical dimensions.

After establishing these foundational building blocks, we will proceed to discuss the classical perspective on generalization. This section covers key ideas such as empirical risk minimization (discussed in further detail in Chapter 2) and bias-variance decomposition. The fundamental insights presented in this chapter are intended to provide a good understanding that will facilitate a smooth transition to the modern perspectives on generalization discussed in the subsequent chapter.

3.1 The Goal of Machine-Learning Models

The primary objective of constructing machine-learning models is to emulate and automate manual processes, thus reducing human intervention and increasing efficiency. For example, tasks such as predicting housing prices or forecasting future temperatures exemplify applications where these models provide significant value. Supervised learning, a major branch of machine

DOI: 10.1201/9781003511601-3

learning, focuses on learning a mapping relationship between input data and target output that has been predetermined through manual annotation.

Mathematically, this mapping is formalized by learning a function

$$f : \mathbb{R}^d \to \mathbb{R},$$

which transforms an input vector $\mathbf{x} \in \mathbb{R}^d$ into a real-valued scalar output $y \in \mathbb{R}$. This function is parameterized by a set of parameters $\boldsymbol{\theta}$ that govern the nature of the mapping. For example, in the case of linear regression, the function is typically expressed as

$$f(\mathbf{x}; \boldsymbol{\theta}) = \theta_0 + \theta_1 x_1 + \theta_2 x_2 + \cdots + \theta_d x_d,$$

where θ_0 is the intercept term and $\theta_1, \theta_2, \ldots, \theta_d$ are the coefficients associated with each input feature.

The structure of this mathematical formulation embodies our inductive bias, which encapsulates prior beliefs about the underlying data-generating process. In linear regression, inductive bias assumes that the relationship between the input features and the output is linear. This simplifying assumption not only makes the model tractable but also reflects a hypothesis that often aligns well with various real-world phenomena.

During the learning process, the model parameters $\boldsymbol{\theta}$ are optimized to minimize the discrepancy between the predicted output and the actual target values. Once the optimization converges, the trained model can be used to predict outputs for new input data $\mathbf{x}_{\text{new}}$ according to the relation

$$\hat{y}_{\text{new}} = f(\mathbf{x}_{\text{new}}; \boldsymbol{\theta}).$$

This automatic derivation and data-driven learning of the mapping from data starkly contrasts with traditional rule-based software engineering, where explicit mapping rules must be manually coded. In fact, this is indeed the case when these model weights are learned and used in production. The ability of supervised learning algorithms to learn this mapping logic directly from data is particularly advantageous in scenarios where the underlying processes are too complex for manual rule identification.

It is essential to recognize that any model represents a simplifying assumption about the underlying data-generating process, with these assumptions often being accompanied by an inherent inductive bias. In the context of linear regression, the assumption of linearity implies that the relationship between inputs and outputs is smooth and continuous, meaning that small changes in input features should result in proportionally small changes in the output. Formally, this concept of smoothness can be mathematically articulated as

$$\lim_{\|\Delta \mathbf{x}\| \to 0} \Delta y = \lim_{\|\Delta \mathbf{x}\| \to 0} [f(\mathbf{x} + \Delta \mathbf{x}; \boldsymbol{\theta}) - f(\mathbf{x}; \boldsymbol{\theta})] = 0.$$

When an inductive bias is appropriately aligned with the true underlying process, it can facilitate a generalization effect, enabling the model to perform well on unseen data. Furthermore, our hypothesis regarding the true function may be enriched by prior knowledge of expected regularities or properties in the data, which can be further leveraged through advanced frameworks such as transfer learning.

3.1.1 Revisiting the model training workflow

To thoroughly examine the intricacies of model development through supervised learning, we now revisit the four key components of the model training workflow introduced in Chapter 1: data, model, cost function, and optimization.

3.1.1.1 Data

In supervised learning, the dataset is comprised of pairs

$$\{(\mathbf{x}^{(i)}, y^{(i)})\}_{i=1}^{n},$$

where each $\mathbf{x}^{(i)} \in \mathbb{R}^d$ represents an input feature vector, and $y^{(i)} \in \mathbb{R}$ denotes the corresponding target value. For notational convenience, the entire dataset is often represented as

$$\mathcal{D} = \{(\mathbf{X}, \mathbf{y})\} = \{(\mathbf{x}^{(1)}, y^{(1)}), (\mathbf{x}^{(2)}, y^{(2)}), \dots, (\mathbf{x}^{(n)}, y^{(n)})\},$$

where $\mathbf{X}$ is the matrix collecting all input vectors and $\mathbf{y}$ is the vector of target values. This structured representation facilitates efficient processing and analysis during model training.

Note that $\mathbf{X}$ can also represent unstructured data such as images and texts. In this case, the unstructured data would be preprocessed into a structured format to facilitate the downstream processing and learning.

3.1.1.2 Model

The model class defines the hypothesis space from which we aim to select the best function f that approximates the underlying relationship between inputs and outputs. In linear regression, the hypothesis space consists of all linear functions parameterized by $\boldsymbol{\theta}$. Specifically, the function is expressed as

$$f(\mathbf{x}; \boldsymbol{\theta}) = \theta_0 + \theta_1 x_1 + \theta_2 x_2 + \dots + \theta_d x_d,$$

which can be more succinctly rewritten in vectorized form as

$$f(\mathbf{x}; \boldsymbol{\theta}) = \boldsymbol{\theta}^\top \mathbf{x},$$

where the augmented input vector is defined as

$$\mathbf{x} = [1, x_1, x_2, \dots, x_d]^\top,$$

thereby incorporating the intercept term θ_0. This formulation represents our inductive bias toward linearity.

3.1.1.3   Cost function

The cost function quantitatively measures the discrepancy between the model's predictions and the actual target values. In the context of linear regression, the mean squared error (MSE) is commonly adopted and is defined as

$$J(\boldsymbol{\theta}) = \frac{1}{n} \sum_{i=1}^{n} \left(f(\mathbf{x}^{(i)}; \boldsymbol{\theta}) - y^{(i)} \right)^2 .$$

which represents the empirical risk using the training set. Minimizing $J(\boldsymbol{\theta})$ during the training process ensures that the model's predictions are as close as possible to the observed data, which produces a good approximation of the underlying data-generating process.

3.1.1.4   Optimization

Optimization involves determining the parameter vector $\boldsymbol{\theta}$ that minimizes the cost function $J(\boldsymbol{\theta})$. For linear regression, one common approach is to use the Normal Equation, which provides an analytical solution:

$$\boldsymbol{\theta} = (\mathbf{X}^\top \mathbf{X})^{-1} \mathbf{X}^\top \mathbf{y},$$

where $\mathbf{X}$ is the design matrix containing all input feature vectors, and $\mathbf{y}$ is the vector of corresponding target values. Alternatively, iterative methods such as gradient descent can be employed to approximate the optimal parameters. In gradient descent, the parameter update is given by

$$\boldsymbol{\theta} := \boldsymbol{\theta} - \alpha \nabla J(\boldsymbol{\theta}),$$

where α represents the user-defined learning rate, and $\nabla J(\boldsymbol{\theta})$ is the gradient of the cost function with respect to $\boldsymbol{\theta}$. This iterative approach is particularly useful when the analytical solution is computationally infeasible or when dealing with very large datasets.

In the next section, we will dive deeper into the data component, exploring its role and significance in the overall supervised learning process.

3.2   The Data

A substantial amount of high-quality training data is the cornerstone of today's effective machine-learning systems. One can think of these data as the knowledge of a teacher who provides the correct answers to a student. For instance, when a model is shown a picture of a cat along with the label "This is a cat," it is tasked to learn to recognize and classify similar images in the future. This data-driven teaching process underlies many of the sophisticated applications we use on a daily basis, such as facial recognition features that unlock smartphones.

As introduced earlier, all machine-learning models begin with a given set of training data. In the context of supervised learning, this training data comprises input-output pairs. Suppose there are n such pairs, and consider the i-th pair denoted by

$$\left(\mathbf{x}^{(i)}, y^{(i)}\right).$$

Each entry in the training set must include both the input $\mathbf{x}^{(i)}$ and the output $y^{(i)}$ to facilitate the learning process. The input vector $\mathbf{x}^{(i)}$ is typically bolded to emphasize that it consists of one or more features. These features serve as the predictors that the model uses to generate a scalar prediction $\hat{y}^{(i)}$ for the i-th observation, according to the model function

$$\hat{y}^{(i)} = f\left(\mathbf{x}^{(i)}; \boldsymbol{\theta}\right),$$

where f is the model function parameterized by the vector $\boldsymbol{\theta}$.

It is worth noting that the input vector $\mathbf{x}^{(i)}$ can be referred to by various names. In machine learning, it is often called a *feature* or *predictor*, while in statistics it might be referred to as a *covariate* or *independent variable*. Similarly, the output $y^{(i)}$ is interchangeably known as the *target*, *label*, *outcome*, or *dependent variable*. We will use these terms interchangeably throughout this chapter.

Furthermore, all the features in $\mathbf{x}^{(i)}$ of a single data entry are collectively referred to as an *observation*. When combined, all observations form the design matrix $\mathbf{X}$. By convention, vectors are represented as column vectors. Thus, given the i-th input-output pair $\left(\mathbf{x}^{(i)}, y^{(i)}\right)$, where $\mathbf{x}^{(i)}$ is a vector of features and $y^{(i)}$ is a scalar outcome, the entire dataset can be denoted as

$$\mathcal{D} = (\mathbf{X}, \mathbf{y}).$$

Here, $\mathbf{X}$ is the design matrix, a two-dimensional array with n rows and p columns (with p representing the number of features), and $\mathbf{y}$ is a column vector with n elements:

$$\mathbf{X} = \begin{bmatrix} \mathbf{x}^{(1)\top} \\ \mathbf{x}^{(2)\top} \\ \vdots \\ \mathbf{x}^{(n)\top} \end{bmatrix} \in \mathbb{R}^{n \times p}, \quad \mathbf{y} = \begin{bmatrix} y^{(1)} \\ y^{(2)} \\ \vdots \\ y^{(n)} \end{bmatrix} \in \mathbb{R}^{n}.$$

This structured format is essential for both theoretical analysis and practical implementation of machine-learning algorithms.

As a concrete example, consider the following Python code, which demonstrates how to create a one-dimensional array (vector) and a two-dimensional array (matrix) using randomly generated numbers from the NumPy library, a widely used package for scientific computing in Python.

```
1   # Widely used package for scientific computing
2   import numpy as np
3
4   # Fix random seed for reproducibility
5   np.random.seed(1)
6
7   # Generate a one-dimensional array of random numbers
8   x = np.random.rand(5)
9   print(x)
10  # Expected Output:
11  # array([0.417022, 0.720324, 0.000114, 0.302333, 0.146756])
12
13  # Generate a two-dimensional array (matrix) of random numbers
14  X = np.random.rand(2, 3)
15  print(X)
16  # Expected Output:
17  # array([[0.092339, 0.186260, 0.345561],
18  #        [0.396767, 0.538817, 0.419195]])
```

Listing 3.1
Creating Arrays with NumPy.

This example illustrates how data can be generated and structured into vectors and matrices, which are amenable to further analysis and use in training machine-learning models.

3.2.1 Sampling from the underlying data distribution

In supervised learning, the training dataset

$$\mathcal{D} = (\mathbf{X}, \mathbf{y})$$

is commonly assumed to be drawn from an unknown underlying data-generating distribution

$$\mathcal{P}(\mathbf{x}, y).$$

This relationship is denoted as

$$(\mathbf{X}, \mathbf{y}) \sim \mathcal{P}(\mathbf{x}, y).$$

In practice, the distribution $\mathcal{P}(\mathbf{x}, y)$ is not directly accessible. If it were known, many learning problems would be considerably simplified. Instead, the dataset $\mathcal{D}$ represents a single realization of observations, which are often assumed to be independently and identically distributed (i.i.d.) samples drawn from $\mathcal{P}(\mathbf{x}, y)$.

When we repeatedly draw samples from the true distribution $\mathcal{P}(\mathbf{x}, y)$ to generate additional datasets, each new training set will likely differ from $\mathcal{D}$. In fact, a commonly used technique called bootstrap is based on randomly sampling from $\mathcal{D}$ to generate additional artificial datasets. This variability is the motivation behind data augmentation techniques, which create additional artificial (or bootstrapped) samples for training purposes. By exposing the

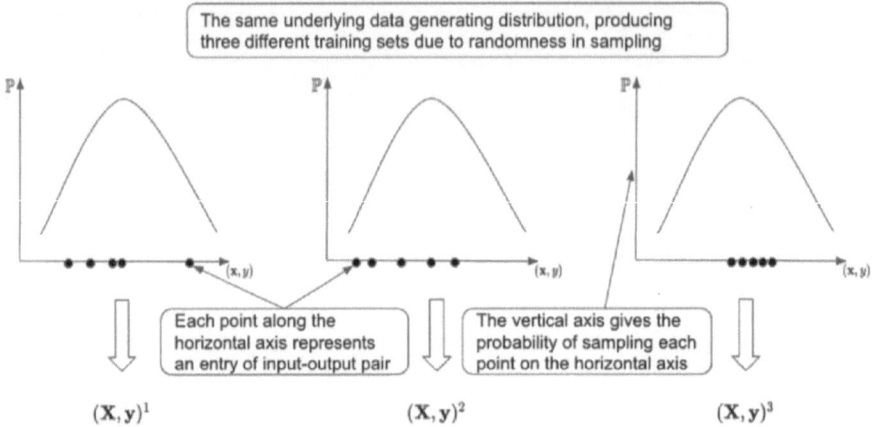

FIGURE 3.1
A simplified process of generating three different training sets based on the same data distribution.

model to a broader range of data variations, these techniques enhance the model's robustness and its ability to generalize to unseen data.

It is important to note, however, that not every additional sample will contribute positively to model learning; some may introduce noise or irrelevant information. Nonetheless, in general, providing the model with diverse training data tend to improve both its training performance and generalization capabilities.

Figure 3.1 illustrates a simplified process of generating three different training sets from the same data distribution. In this example, each training set comprises five different input-output pairs randomly sampled from $\mathcal{P}(\mathbf{x}, y)$. Due to the randomness inherent in the sampling process, and potential noise perturbations, these training sets will most likely differ from one another.

It is evident from this process that the available training set is merely one snapshot among many possible realizations. Sampling another training set from $\mathcal{P}(\mathbf{x}, y)$ would likely result in a different collection of observations. Moreover, both the input matrix $\mathbf{X}$ and the output vector $\mathbf{y}$ can contain either continuous or discrete variables. In our discussion, we focus on linear regression, which typically assumes a continuous target variable y.

To illustrate the concept of sampling variability, consider the following Python code. Previously, we used the function `np.random.rand()` to generate random numbers uniformly distributed between 0 and 1. This function ensures that every number within the interval $[0, 1]$ is equally likely to be sampled. However, it is possible to modify the range or employ different probability distributions, such as the normal distribution, which is characterized by its mean and variance. The code listing below demonstrates these differences.

```
1  # Draw random samples from a uniform distribution ranging from -1
        to 0
2  x = np.random.uniform(-1, 0, 5)
3  print(x)
4  # Expected Output:
5  # array([-0.3147805 , -0.79554775, -0.12188256, -0.97261241,
        -0.32953249])
6
7  # Drawing again will most likely result in a different set of
        numbers
8  x = np.random.uniform(-1, 0, 5)
9  print(x)
10 # Expected Output:
11 # array([-0.5826952 , -0.44131017, -0.85961306, -0.80189851,
        -0.19925543])
12
13 # Change to a normal distribution with mean 1 and standard
        deviation 2
14 x = np.random.normal(1, 2, 5)
15 print(x)
16 # Expected Output:
17 # array([2.45995119, 1.74598758, 2.06762182, 0.8160534 ,
        4.82764078])
```

Listing 3.2
Demonstrating Sampling Variability with Different Distributions.

As observed from the above examples, each execution generates a different set of numbers. Drawing a list of five numbers offers infinitely many possibilities. On one hand, we lack direct access to the true underlying distribution, which might be uniform over a specific range, normal with a particular mean and variance, or even a more complex distribution defined by other parameters. On the other hand, even if the true data-generating distribution were known, each sampling process would still yield different results due to the inherent randomness of the sampling mechanism.

Formally, if an individual input $\mathbf{x}^{(i)}$ is drawn from the marginal distribution $\mathcal{P}(\mathbf{x})$, then each $\mathbf{x}^{(i)}$ is often assumed to be an i.i.d. sample:

$$\mathbf{x}^{(i)} \sim \mathcal{P}(\mathbf{x}), \quad \text{for } i = 1, 2, \ldots, n.$$

This i.i.d. assumption is fundamental, as it underpins many theoretical guarantees in machine learning and is crucial for ensuring that the model generalizes well to new, unseen data.

Understanding and managing the variability in sampling is vital for model training. Even with a simple model like linear regression, the capacity to generate diverse training datasets—through techniques such as data augmentation—can significantly enhance the model's resilience to fluctuations and noise encountered in real-world scenarios.

In summary, the quality, structure, and inherent variability of the training data play a critical role in determining the performance and generalization capabilities of machine-learning models. A thorough understanding of the data sampling process, along with the careful use of techniques to increase data diversity, is a good step toward building robust and effective models.

3.2.2 The train-test split

As discussed in the previous chapters, the primary objective of any machine-learning model is to achieve strong predictive performance on future, unseen datasets. In practice, these unseen datasets are commonly referred to as the *test set*. However, during the model development phase, the true test set is not available, and we must rely solely on the available dataset

$$\mathcal{D} = (\mathbf{X}, \mathbf{y})$$

to evaluate and refine our model. To rigorously assess the model's predictive capabilities and to mitigate potential issues such as overfitting and selection bias, it is standard practice to partition $\mathcal{D}$ into distinct, non-overlapping subsets: a training set, a validation set, and a test set.

The test set, denoted by

$$\mathcal{D}^{\text{test}} = (\mathbf{X}^{\text{test}}, \mathbf{y}^{\text{test}}),$$

serves as a proxy for evaluating the model's generalization performance on future data. Two key principles govern the use of the test set:

1. Exclusion from Training: The test set must remain entirely separate from the model training process. This means that none of the data in $\mathcal{D}^{\text{test}}$ should be used for training or hyperparameter tuning. This strict separation prevents data leakage, ensuring that the model does not inadvertently learn from the test set and thereby maintains unbiased performance estimates. For example, the test set in a time series often comes from the last periods of the data instead of earlier ones to avoid looking into the future.

2. Proxy for Future Performance: The test set functions as a stand-in for future, unseen data, providing an unbiased estimate of how well the model is expected to perform on new data drawn from the same underlying distribution $\mathcal{P}(\mathbf{x}, y)$. Although the test set is intended to reflect future performance, it is important to acknowledge that it may not capture all possible variations due to factors such as data drift, where the distribution of incoming data may change over time.

Once the test set $\mathcal{D}^{\text{test}}$ has been allocated, the remaining data

$$\mathcal{D}^{\text{train-val}} = (\mathbf{X}^{\text{train-val}}, \mathbf{y}^{\text{train-val}})$$

is typically further subdivided into a training set and a validation set. The training set, denoted as

$$\mathcal{D}^{\text{train}} = (\mathbf{X}^{\text{train}}, \mathbf{y}^{\text{train}}),$$

is used exclusively for model fitting and parameter estimation, while the validation set,

$$\mathcal{D}^{\text{val}} = (\mathbf{X}^{\text{val}}, \mathbf{y}^{\text{val}}),$$

is utilized for hyperparameter tuning, model selection, and performance monitoring during the training process. It also serves as a proxy to the future test set but is available for model selection purpose.

Formally, the partitioning of the dataset can be expressed as:

$$\mathcal{D}^{\text{train}} \cup \mathcal{D}^{\text{val}} \cup \mathcal{D}^{\text{test}} = \mathcal{D},$$

with the subsets satisfying the mutual exclusivity condition:

$$\mathcal{D}^{\text{train}} \cap \mathcal{D}^{\text{val}} = \mathcal{D}^{\text{train}} \cap \mathcal{D}^{\text{test}} = \mathcal{D}^{\text{val}} \cap \mathcal{D}^{\text{test}} = \emptyset.$$

This careful and deliberate splitting of the dataset is crucial for obtaining a reliable estimate of the model's performance on future data. By strictly separating training and evaluation processes, we ensure that the assessment of generalization capability is both fair and robust, ultimately leading to more trustworthy and effective machine-learning models.

3.2.3 Selection bias and cross-validation

There exists an inherent risk of selection bias when partitioning a dataset into training, validation, and test sets. The specific manner in which the dataset is split may inadvertently favor certain patterns in the data, leading to models that perform well on the validation and test sets by chance rather than by virtue of true generalization. Such bias can result in overly optimistic performance estimates if not properly addressed.

To mitigate this risk, cross-validation techniques are widely employed. Cross-validation systematically partitions the data into multiple distinct training and validation sets, ensuring that the model's performance is evaluated consistently across various subsets. This approach reduces the likelihood of overfitting to any particular data split and yields a more robust and reliable estimate of the model's generalization capability on unseen data.

The proportion of data allocated to each subset often depends on the overall size and characteristics of the dataset. For smaller datasets, a common heuristic is to allocate approximately 60% for training, 20% for validation, and 20% for testing (i.e., a 6:2:2 split). Conversely, for larger datasets, a typical split might be 80% for training, 10% for validation, and 10% for testing (i.e., an 8:1:1 ratio). These ratios are merely guidelines and can be adjusted to suit the specific requirements of the task and the inherent variability of the data.

Figure 3.2 illustrates three different methods for splitting the same dataset into training, validation, and test sets. In this setup, the training set is used to optimize the model parameters, the validation set assists in hyperparameter tuning and model selection, and the test set serves as an (hopefully) unbiased benchmark for evaluating the model's generalization performance.

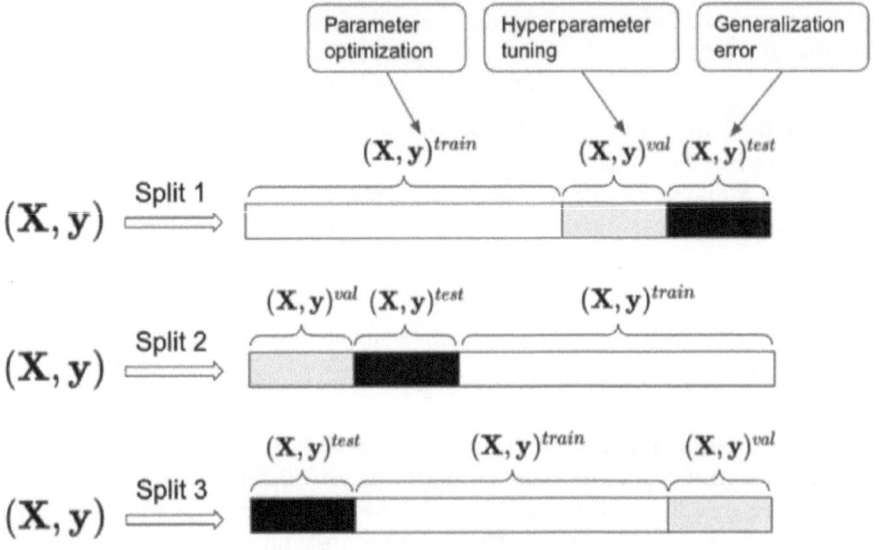

FIGURE 3.2
Three different ways of splitting the same dataset into training, validation, and test set.

To further illustrate the train-test split procedure, consider a toy dataset. The following Python code snippet creates a simple set of input features and corresponding target values and then splits the data into training and test sets according to a predefined ratio.

```python
import random
import numpy as np
import matplotlib.pyplot as plt

# Reshape to a column representation; -1 infers the number of
    rows automatically
X = np.array(range(10)).reshape(-1, 1)  # Input features
y = np.array([1.5, 2.5, 3.7, 3.8, 5.1, 5.8, 6.2, 5.6, 5.2, 5.3]).
    reshape(-1, 1)

# Define the split ratio for the training set
train_ratio = 0.8

# Randomly sample indices for the training set
train_idx = random.sample(range(len(X)), int(train_ratio * len(X)
    ))

# Allocate training set based on selected indices
X_train = X[train_idx]
y_train = y[train_idx]

# Allocate test set based on indices not included in the training
    set
```

```
20  X_test = X[[idx for idx in range(len(X)) if idx not in train_idx
    ]]
21  y_test = y[[idx for idx in range(len(X)) if idx not in train_idx
    ]]
```

Listing 3.3
Train-Test Split Using a Toy Dataset.

In this example, the dataset **X** consists of integers from 0 to 9 reshaped into a column vector, while **y** contains the corresponding target values. An 80% training split is implemented by randomly selecting indices via Python's `random.sample()` function. The remaining indices form the test set. Although the validation set is not explicitly shown in this code, it is instrumental for hyperparameter tuning.

It is important to note that while the validation set is critical for model tuning, there is a risk of overfitting to this set if the model becomes overly specialized to its specific characteristics. To address this concern, cross-validation methods such as k-fold cross-validation are employed. In k-fold cross-validation, the dataset is divided into k equally sized folds. The model is then trained k times, each time using $k-1$ folds for training and the remaining fold for validation. The average performance across all k iterations provides a robust estimate of the model's generalization ability.

Furthermore, nested cross-validation extends this concept by incorporating an inner loop for hyperparameter tuning within each outer fold. This approach prevents bias in the performance estimates due to the hyperparameter selection process, resulting in an even more accurate assessment of model generalization.

Platforms such as Kaggle try to mitigate the risk of selection bias by withholding the test set until the final evaluation phase, thereby preventing participants from tuning their models based on exposed test set performance, which is displayed in the public leaderboard. This practice is crucial in avoiding data leakage and ensuring a fair assessment of the model's true generalization capabilities.

In summary, careful management of selection bias through proper dataset partitioning and the use of cross-validation techniques is essential to obtain reliable performance estimates. These practices not only help prevent overfitting but also ensure that the model's performance on future, unseen data is accurately evaluated.

3.3 The Model

In the context of linear regression, the model is defined by both a set of parameters and an architectural structure that collectively determine how the input features interact to produce predictions. The parameters, often referred to as *weights*, are scalar values that are adjusted during the training process to optimize the model's performance. Our primary goal is to determine the optimal set of weights that best maps the input features $\mathbf{X}$ to the target values $\mathbf{y}$.

Let

$$\mathbf{w} = [w_1, w_2, \ldots, w_p]^\top$$

denote the weight vector corresponding to the p input features in $\mathbf{X}$. Each weight w_j modulates the influence of the corresponding feature x_j on the prediction. In essence, w_j quantifies the contribution of x_j to the final prediction output $\hat{y}$. The objective is to find the optimal set of weights $\mathbf{w}^*$ that minimizes a predefined cost function, typically the MSE, ensuring that the model predictions closely approximate the true target values.

The architecture of the model dictates the functional relationship between the input features and the weights. In linear regression, we typically assume the following linear generative process:

$$y = f(\mathbf{X}; \mathbf{w}) + \epsilon,$$

where the function

$$f(\mathbf{X}; \mathbf{w}) = \mathbf{X}\mathbf{w}$$

represents the deterministic component of the model. Here, $\mathbf{X}$ is the matrix of input features, and $\mathbf{w}$ is the vector of weights. The term ϵ denotes additive random noise, which is commonly assumed to be uncorrelated with $\mathbf{X}$ and normally distributed with zero mean and a fixed variance σ^2, that is, $\epsilon \sim \mathcal{N}(0, \sigma^2)$.

For a single observation, the model's prediction is given by:

$$\hat{y}^{(i)} = \mathbf{x}^{(i)\top}\mathbf{w},$$

where $\mathbf{x}^{(i)}$ represents the i-th row of $\mathbf{X}$ (i.e., the feature vector corresponding to the i-th observation). This formulation explicitly shows that the prediction is a weighted sum of the input features.

To recap, the model architecture serves as a prediction function that maps the input features $\mathbf{X}$ and weights $\mathbf{w}$ to the predicted outputs $\hat{\mathbf{y}}$. Using compact matrix notation, this relationship is succinctly expressed as:

$$\hat{\mathbf{y}} = f(\mathbf{X}; \mathbf{w}) = \mathbf{X}\mathbf{w}.$$

This is a parametric model because the function f is completely parameterized by the weight vector $\mathbf{w}$. While the input $\mathbf{X}$ is fixed based on the training data, $\mathbf{w}$ represents the tunable parameters that are iteratively optimized during the training process. Furthermore, this framework not only underpins linear regression but also provides the foundation for more complex models where additional layers and nonlinearities are introduced to capture intricate patterns within the data.

3.3.1 Parametric versus non-parametric models

A *parametric model* is characterized by a fixed and typically small number of parameters that do not grow with the size of the dataset. In such models, the functional form of the relationship between the input features and the target variable is predetermined. Linear regression is a canonical example of a parametric model due to its explicit linear structure and limited parameter count. In this framework, the model assumes that the output is a linear combination of the input features, and the number of parameters remains constant regardless of the amount of available data.

In contrast, a *non-parametric model* does not assume a fixed form for the function f that maps inputs to outputs. Instead, it is capable of adapting its complexity to the underlying data, potentially employing an unbounded number of parameters as more data become available. This flexibility allows non-parametric models to capture complex and nonlinear relationships inherent in the data. For instance, deep neural networks are often considered non-parametric due to their vast number of parameters and their ability to learn intricate representations, even though they technically adhere to a defined architecture.

It is also instructive to review the matrix multiplication process, which plays a central role in the formulation of linear regression. Recall that a matrix is defined by its dimensions, where a matrix with shape (n, p) has n rows and p columns. Given two matrices $\mathbf{A} \in \mathbb{R}^{n \times p}$ and $\mathbf{B} \in \mathbb{R}^{q \times k}$, the product $\mathbf{AB}$ is defined only if the inner dimensions match (i.e., $p = q$). The resulting matrix $\mathbf{C} = \mathbf{AB}$ will then have the shape (n, k).

In linear regression, the prediction is obtained by multiplying the design matrix $\mathbf{X}$ by the weight vector $\mathbf{w}$. This operation is succinctly represented as:

$$\hat{\mathbf{y}} = \mathbf{X}\mathbf{w},$$

where $\mathbf{X} \in \mathbb{R}^{n \times p}$ contains the input features, and $\mathbf{w} \in \mathbb{R}^p$ contains the corresponding weights. This formulation emphasizes the parametric nature of linear regression, as the mapping of inputs to outputs is entirely determined by the finite set of parameters in $\mathbf{w}$.

Overall, the distinction between parametric and non-parametric models is important in understanding model flexibility and capacity. Parametric models offer simplicity and interpretability with a fixed number of parameters, while non-parametric models provide greater adaptability to complex patterns in the data. Recognizing these differences is crucial when selecting an appropriate model for a specific task, as it directly influences the model's ability to generalize to new, unseen data.

3.3.2 The bias trick

In linear regression, a bias term w_0 is often incorporated into the model to account for the intercept, enabling the model to fit data that do not necessarily pass through the origin. Including this bias term modifies the prediction equation to:

$$\hat{\mathbf{y}} = w_0 + \mathbf{X}\mathbf{w},$$

where $\mathbf{X}$ represents the design matrix containing the input features and $\mathbf{w}$ is the weight vector associated with these features.

A common technique to simplify the model representation and computation is known as the *bias trick*. This method involves augmenting the design matrix $\mathbf{X}$ by adding an additional column of ones. By doing so, the bias term w_0 is absorbed into the weight vector, allowing the prediction to be expressed as a single matrix-vector multiplication. Specifically, the augmented design matrix, denoted by $\mathbf{X}_{\text{new}}$, is defined as:

$$\mathbf{X}_{\text{new}} = \begin{bmatrix} 1 & x_1^{(1)} & x_2^{(1)} & \cdots & x_p^{(1)} \\ 1 & x_1^{(2)} & x_2^{(2)} & \cdots & x_p^{(2)} \\ \vdots & \vdots & \vdots & \ddots & \vdots \\ 1 & x_1^{(n)} & x_2^{(n)} & \cdots & x_p^{(n)} \end{bmatrix},$$

where $x_j^{(i)}$ denotes the value of the j-th feature for the i-th observation. Correspondingly, the weight vector is augmented to include the bias term:

$$\mathbf{w}_{\text{new}} = \begin{bmatrix} w_0 \\ w_1 \\ w_2 \\ \vdots \\ w_p \end{bmatrix}.$$

Thus, the prediction function can now be succinctly written as:

$$\hat{\mathbf{y}} = \mathbf{X}_{\text{new}}\mathbf{w}_{\text{new}}.$$

To implement this prediction function, one begins by augmenting the design matrix $\mathbf{X}$ with a column of ones. The weight vector $\mathbf{w}$ (which now includes the bias term) is typically initialized with random values, often drawn from a standard normal distribution. The predicted values are then computed via matrix-vector multiplication. For example, the following Python code demonstrates this process:

```
1   import numpy as np
2   import matplotlib.pyplot as plt
3
4   # Augment the design matrix by adding a column of ones to account
        for the bias term
5   X_new = np.hstack((np.ones_like(X), X))
6
7   # Initialize a random set of weights, including the bias term
8   w = np.random.normal(0, 1, (X_new.shape[1], 1))
9
```

```
10  # Compute the predictions via matrix-vector multiplication
11  f = X_new @ w
```

Listing 3.4
Implementation of the Bias Trick in Python.

In the code above, the augmented matrix X_{new} now contains an extra column that represents the bias term, while w includes both w_0 and the other weights $w_1, w_2, \ldots, w_p$. For instance, if the model has only one feature (apart from the bias), the weight vector will have two parameters: the intercept w_0 and the slope w_1. Consequently, the model prediction becomes a straight line determined by these two parameters.

To further illustrate this concept, consider the following snippet that visualizes the initial, unoptimized model by plotting the training data along with the prediction line based on the randomly initialized weights:

```
1   # Create a list of equally spaced input values for plotting
2   X_vals = np.linspace(X.min(), X.max(), 100)[:, np.newaxis]
3
4   # Augment the input values with a column of ones
5   X_vals_new = np.hstack((np.ones_like(X_vals), X_vals))
6
7   # Generate predictions using the current weights
8   y_vals = X_vals_new @ w
9
10  # Plot the training data and the model's prediction line
11  fig, ax = plt.subplots()
12  ax.scatter(X, y, c='black')
13  ax.set_xlabel('x')
14  ax.set_ylabel('y')
15  ax.plot(X_vals, y_vals, linewidth=2)
16  plt.show()
```

Listing 3.5
Visualizing the Initial Model.

As shown in the resulting plot (referenced in Figure 3.3), the initial model is represented by a straight line based on the randomly initialized weights. At this stage, the line does not fit the training data well. However, in subsequent steps, the model's weights will be optimized—either through closed-form solution (called normal equation for linear regression) or iterative methods such as gradient descent—to minimize a cost function, thereby improving the fit to the training data and enhancing generalization to unseen data.

3.4   The Cost Function

Almost every machine-learning algorithm relies on a feedback mechanism to assess the quality of its predictions and guide subsequent updates. This

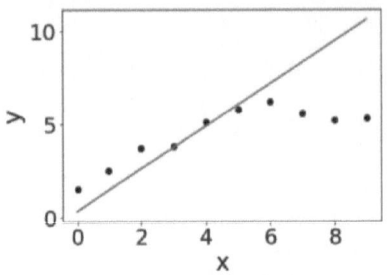

FIGURE 3.3
The initial model as represented by the straight line.

feedback is provided by the cost function, which quantifies the error introduced by the current model. In essence, the cost function evaluates how closely
the model's predictions align with the actual target values, thereby serving as
a critical tool for iterative model improvement and parameter optimization.

In the context of linear regression, the cost function measures the goodness of fit of the model to the training data. For a given input-output pair
$(\mathbf{x}^{(i)}, y^{(i)})$, where $\mathbf{x}^{(i)} \in \mathbb{R}^d$ represents the input feature vector and $y^{(i)} \in \mathbb{R}$ is
the corresponding target value, the model's prediction can be denoted as

$$\hat{y}^{(i)} = f(\mathbf{x}^{(i)}; \mathbf{w}).$$

The *residual* for the i-th data point is then defined as

$$r^{(i)} = y^{(i)} - \hat{y}^{(i)} = y^{(i)} - f(\mathbf{x}^{(i)}; \mathbf{w}).$$

These residuals represent the vertical distances between the model's predictions and the actual data points, serving as indicators of prediction errors.

A widely used method to aggregate these residuals is the *Sum of Squared
Errors (SSE)*, defined as:

$$\mathrm{SSE}(\mathbf{w}) = \sum_{i=1}^{n} \left(y^{(i)} - f(\mathbf{x}^{(i)}; \mathbf{w}) \right)^2 = \sum_{i=1}^{n} \left(r^{(i)} \right)^2.$$

The SSE has several desirable properties: it is nonnegative, differentiable, and,
by squaring the residuals, it disproportionately penalizes larger errors. This
characteristic encourages the model to focus on reducing significant discrep-
ancies between predictions and actual values.

An alternative to SSE is the *MSE*, which normalizes the SSE by dividing
by the number of observations:

$$\mathrm{MSE}(\mathbf{w}) = \frac{1}{n} \sum_{i=1}^{n} \left(y^{(i)} - f(\mathbf{x}^{(i)}; \mathbf{w}) \right)^2 = \frac{1}{n} \mathrm{SSE}(\mathbf{w}).$$

While both SSE and MSE yield the same optimal weight estimates, MSE is
often preferred as it provides a more interpretable measure of the average
prediction error.

Both SSE and MSE are instances of the cost function $Q(\mathbf{w})$, which is defined solely as a function of the model's weights $\mathbf{w}$, independent of the fixed design matrix $\mathbf{X}$. Our goal is thus to find the weights $\mathbf{w}$ that correspond to the smallest $Q(\mathbf{w})$.

This cost function can also be elegantly expressed using linear algebra. Let $\boldsymbol{\Phi}$ denote the design matrix augmented with a column of ones to incorporate the bias term:

$$\boldsymbol{\Phi} = \begin{bmatrix} 1 & \mathbf{x}^{(1)\top} \\ 1 & \mathbf{x}^{(2)\top} \\ \vdots & \vdots \\ 1 & \mathbf{x}^{(n)\top} \end{bmatrix} \in \mathbb{R}^{n \times (p+1)},$$

and let the augmented weight vector be

$$\mathbf{w} = \begin{bmatrix} w_0 \\ w_1 \\ w_2 \\ \vdots \\ w_p \end{bmatrix} \in \mathbb{R}^{(p+1) \times 1}.$$

With these definitions, the predictions for all n observations can be succinctly written as:

$$\hat{\mathbf{y}} = \boldsymbol{\Phi}\mathbf{w}.$$

Thus, the cost function takes the form:

$$Q(\mathbf{w}) = (\mathbf{y} - \boldsymbol{\Phi}\mathbf{w})^\top (\mathbf{y} - \boldsymbol{\Phi}\mathbf{w}) = \mathbf{y}^\top\mathbf{y} - 2\mathbf{y}^\top\boldsymbol{\Phi}\mathbf{w} + \mathbf{w}^\top\boldsymbol{\Phi}^\top\boldsymbol{\Phi}\mathbf{w}.$$

This quadratic form is particularly advantageous for optimization because its smoothness and differentiability allow for the derivation of closed-form solutions (e.g., via the Normal Equation) and the efficient application of iterative methods such as gradient descent.

To implement the cost function in Python, one typically calculates the residuals and then computes their sum of squares using NumPy's dot product functionality. For example:

```python
import numpy as np

# Calculate the residuals between actual target values y and
    predictions f
resid = y - f

# Compute the total cost as the dot product of the residual
    vector with itself
Q = np.dot(resid.T, resid)
print(Q)
# Expected Output (example):
# array([[58.46752305]])
```

Listing 3.6
Implementing the Cost Function in Python.

In this snippet, the residuals are computed by subtracting the predicted values $\hat{\mathbf{y}}$ from the actual target values $\mathbf{y}$. The total cost Q is then obtained by taking the dot product of the residual vector with itself, effectively summing the squared residuals. This implementation highlights the practical utility of the cost function in quantifying model error and guiding the optimization process.

3.5 The Optimization Algorithm

Optimization is a fundamental component of all learning algorithms, serving as the central mechanism by which models iteratively improve their performance. In essence, optimization involves adjusting the model parameters to minimize the cost function, thereby reducing prediction errors and enhancing the model's overall accuracy.

The central goal of optimization in machine learning is to determine the optimal set of parameters that minimizes the cost function. Formally, given a cost function $Q(\mathbf{w})$ defined over the model's weight vector $\mathbf{w}$, the optimal weight vector $\mathbf{w}^*$ is given by

$$\mathbf{w}^* = \arg\min_{\mathbf{w}} Q(\mathbf{w}).$$

This expression says that $\mathbf{w}^*$ is the point at which the cost function attains its minimum value. At this minimum, the derivative of the cost function with respect to the weights is zero, indicating that the function is at a stationary point.

In the case of multiple features, reaching a stationary point implies that the gradient vector of the cost function with respect to all weights vanishes. The gradient vector, denoted by $\nabla_{\mathbf{w}} Q$, comprises the partial derivatives of Q with respect to each individual weight w_j:

$$\nabla_{\mathbf{w}} Q = \begin{bmatrix} \frac{\partial Q}{\partial w_1} \\ \frac{\partial Q}{\partial w_2} \\ \vdots \\ \frac{\partial Q}{\partial w_p} \end{bmatrix} = \mathbf{0}.$$

For linear regression, we know that the cost function can be expressed in the quadratic form

$$Q(\mathbf{w}) = \mathbf{y}^\top \mathbf{y} - 2\mathbf{y}^\top \mathbf{\Phi} \mathbf{w} + \mathbf{w}^\top \mathbf{\Phi}^\top \mathbf{\Phi} \mathbf{w},$$

where $\mathbf{\Phi}$ is the design matrix (often augmented to include a column of ones for the bias term) and $\mathbf{y}$ is the vector of target values.

A key component of this cost function is the quadratic term $\mathbf{w}^\top \mathbf{\Phi}^\top \mathbf{\Phi} \mathbf{w}$. Differentiating this term with respect to $\mathbf{w}$ using the chain rule and recognizing that $\mathbf{\Phi}^\top \mathbf{\Phi}$ is symmetric, we obtain:

$$\frac{\partial}{\partial \mathbf{w}} \left(\mathbf{w}^\top \mathbf{\Phi}^\top \mathbf{\Phi} \mathbf{w} \right) = 2\mathbf{\Phi}^\top \mathbf{\Phi} \mathbf{w}.$$

Taking into account the entire cost function, the gradient of $Q(\mathbf{w})$ with respect to $\mathbf{w}$ is therefore given by:

$$\nabla_{\mathbf{w}} Q = -2\mathbf{\Phi}^\top \mathbf{y} + 2\mathbf{\Phi}^\top \mathbf{\Phi} \mathbf{w}.$$

Setting this gradient equal to zero to locate the stationary point yields:

$$-2\mathbf{\Phi}^\top \mathbf{y} + 2\mathbf{\Phi}^\top \mathbf{\Phi} \mathbf{w} = \mathbf{0}.$$

Dividing through by 2 simplifies the equation to:

$$\mathbf{\Phi}^\top \mathbf{\Phi} \mathbf{w} = \mathbf{\Phi}^\top \mathbf{y}.$$

This expression is known as the *Normal Equation*, which provides a closed-form solution for the optimal weights:

$$\mathbf{w}^* = \left(\mathbf{\Phi}^\top \mathbf{\Phi} \right)^{-1} \mathbf{\Phi}^\top \mathbf{y},$$

assuming that $\mathbf{\Phi}^\top \mathbf{\Phi}$ is invertible.

In addition, it is important to note that the cost function $Q(\mathbf{w})$ for linear regression is convex. This convexity guarantees that any stationary point is a global minimum, thereby simplifying the optimization process. In practice, however, iterative methods such as gradient descent are frequently employed, especially when dealing with large datasets where computing the closed-form solution may be computationally expensive.

To sum up, the optimization algorithm in linear regression seeks to find the weight vector $\mathbf{w}^*$ that minimizes the cost function $Q(\mathbf{w})$. By setting the gradient of the cost function to zero and solving the resulting Normal Equation, one obtains a closed-form solution that forms the backbone of the learning process, enabling the model to make increasingly accurate predictions.

3.5.1 Multiple minima and convexity

In optimization, the goal is to find the global minimum of the cost function $Q(\mathbf{w})$. However, the landscape of $Q(\mathbf{w})$ may contain several stationary points where the gradient is zero. When $\nabla Q(\mathbf{w}) = \mathbf{0}$, one of the following scenarios can occur:

1. Global Optimum: This is the unique point where the cost function reaches its absolute minimum value. In convex functions, the global optimum is the only stationary point, ensuring that any algorithm converging to a zero gradient has found the best possible solution.

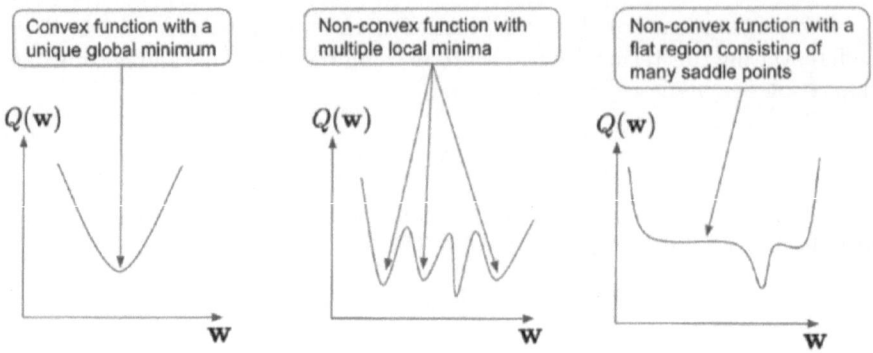

FIGURE 3.4
Three different scenarios when locating the global minimum.

2. Local Optimum: In non-convex functions, multiple local minima can exist. A local optimum is a point where the cost function has a minimum value within a localized region, yet it may not be the lowest value over the entire domain. Distinguishing between a local and a global minimum in such cases can be challenging.

3. Saddle Point: A saddle point is a stationary point where the gradient vanishes, but the point is neither a true minimum nor a maximum. Instead, the function exhibits mixed curvature—acting like a minimum in some directions and a maximum in others. Saddle points can mislead optimization algorithms, particularly those based on gradient descent, into halting progress toward the global optimum.

Figure 3.4 illustrates these three scenarios. In the left panel, a convex function is depicted where the only stationary point is the global minimum. The middle panel shows a non-convex function with multiple local minima, while the right panel illustrates a flat region containing several saddle points. Although all these points satisfy $\nabla Q(\mathbf{w}) = \mathbf{0}$, only the global optimum represents the absolute lowest cost.

In linear regression, the cost function $Q(\mathbf{w})$ is convex due to the quadratic nature of its formulation. This convexity guarantees that the stationary point obtained from the Normal Equation is the unique global minimum. Recall that the Normal Equation is given by:

$$\mathbf{\Phi}^\top \mathbf{\Phi} \mathbf{w} = \mathbf{\Phi}^\top \mathbf{y},$$

which leads to the closed-form solution:

$$\mathbf{w}^* = \left(\mathbf{\Phi}^\top \mathbf{\Phi}\right)^{-1} \mathbf{\Phi}^\top \mathbf{y}.$$

Furthermore, the Hessian matrix of the cost function is:

$$\nabla^2 Q(\mathbf{w}) = 2\mathbf{\Phi}^\top \mathbf{\Phi}.$$

Since $\mathbf{\Phi}^\top\mathbf{\Phi}$ is positive definite, the Hessian is also positive definite, confirming that the curvature of $Q(\mathbf{w})$ is uniformly upward. This positive curvature ensures that the stationary point is indeed a unique global minimum.

Model training generally involves iteratively updating the weight vector using algorithms such as gradient descent. However, when a closed-form solution is available, as in linear regression, the optimization process reduces to a single calculation, obviating the need for iterative updates. In many real-world applications, though, especially with complex and non-convex cost functions, closed-form solutions are not feasible. In such cases, iterative methods become indispensable.

The following Python implementation demonstrates how this closed-form solution is applied:

```python
import numpy as np

# Solve for the optimal weight vector using the Normal Equation
w_star = np.linalg.solve(X_new.T @ X_new, X_new.T @ y)

# Generate model predictions with the optimal weights
f_star = X_new @ w_star

# Calculate the residuals with the optimal weights
resid = y - f_star

# Calculate the cost with the optimal weights
Q = np.dot(resid.T, resid)
print(Q)
# Output:
# array([[6.33963636]])
```

Listing 3.7
Computing the Optimal Weights Using the Normal Equation.

In this code, we first compute the optimal weight vector $\mathbf{w}^*$ by solving the Normal Equation. We then generate the corresponding predictions $\hat{\mathbf{y}}^*$ by multiplying the augmented design matrix X_{new} with $\mathbf{w}^*$. The residuals are calculated by subtracting these predictions from the actual target values, and finally, the cost Q is determined as the dot product of the residual vector with itself. Notably, the cost Q decreases significantly from its initial value, indicating that the model's performance has improved after optimization.

To further illustrate the effect of this optimization, we can visualize the fitted line. The following code snippet plots the training data along with the regression line obtained using the optimal weights:

```python
import matplotlib.pyplot as plt

# Create a range of input values for plotting the fitted line
X_vals = np.linspace(X.min(), X.max(), 100)[:, np.newaxis]

# Add a column of ones to the input values for the bias term
X_vals_new = np.hstack((np.ones_like(X_vals), X_vals))
```

```
 9  # Generate predictions using the optimal weights
10  y_vals = X_vals_new @ w_star
11
12  # Plot the training data and the fitted line
13  fig, ax = plt.subplots()
14  ax.scatter(X, y, c='black')
15  ax.set_xlabel('x')
16  ax.set_ylabel('y')
17  ax.plot(X_vals, y_vals, linewidth=2)
18  plt.show()
```

Listing 3.8
Visualizing the Fitted Line Using Optimal Weights.

Figure 3.5 displays the linear regression fit based on the closed-form solution, clearly demonstrating a close alignment between the fitted line and the training data points.

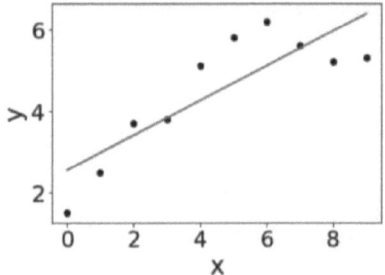

FIGURE 3.5
Linear regression fit using the closed-form solution.

Overall, the convex nature of the cost function in linear regression guarantees the uniqueness of the global minimum, which can be efficiently obtained using the closed-form solution provided by the Normal Equation. However, it is important to acknowledge that the computation of the closed-form solution may become expensive in practice. In particular, the matrix $\Phi^\top \Phi$ can be very large and dense, rendering the storage and inversion of this matrix computationally intensive or even infeasible in certain scenarios.

Understanding the gradient, the nature of stationary points, and the conditions for convexity are crucial to understanding how optimization algorithms navigate the cost landscape. While closed-form solutions are both elegant and effective for linear models, many complex models lack such solutions and, therefore, rely on iterative and computationally efficient optimization techniques such as gradient descent.

3.5.2 The gradient descent algorithm

Gradient descent is one of the most fundamental optimization algorithms in machine learning, playing a critical role in the minimization of cost functions.

It is widely adopted because it can iteratively adjust the model's parameters (commonly referred to as weights) in the direction that most effectively reduces the cost, thereby incrementally improving the model's predictive performance. Even though this algorithm may sometimes converge to local minima or become trapped at saddle points when dealing with non-convex cost functions, it remains the most prevalent optimization technique—especially in the domain of deep learning.

The operation of gradient descent is based on the principle of using the gradient of the cost function as a guide to navigate the parameter space toward a minimum. Specifically, consider a differentiable cost function $Q(\mathbf{w})$ with respect to the weight vector $\mathbf{w}$. The gradient descent algorithm updates the weight vector by moving it in the opposite direction of the gradient at the current point. This process can be expressed by the following update rule:

$$\mathbf{w} \leftarrow \mathbf{w} - \eta \nabla Q(\mathbf{w}),$$

where η is the learning rate, a small positive scalar that determines the step size for each update, and $\nabla Q(\mathbf{w})$ denotes the gradient of the cost function evaluated at $\mathbf{w}$.

The rationale behind this update rule is that the gradient vector points in the direction of the steepest ascent on the cost surface. By moving in the opposite direction, the algorithm ensures that the cost function decreases most rapidly. The learning rate η is an important hyperparameter because it controls the magnitude of each step: if η is too large, the updates may overshoot the minimum, while if it is too small, the convergence process becomes excessively slow.

When visualizing the cost function $Q(\mathbf{w})$ as a function of the weights, one can observe that the gradient vector at any point points toward the steepest increase in cost. Consequently, by taking steps in the reverse direction of this gradient, the gradient descent algorithm efficiently navigates the parameter space toward a local or global minimum. This simple yet powerful mechanism is at the heart of many modern machine-learning algorithms.

Figure 3.6 illustrates the gradual descent update process in a univariate linear regression scenario, where the cost function is dependent on a single weight parameter. In this visualization, various starting points are shown, each demonstrating how the algorithm iteratively converges toward the minimum. With every iteration, the weight parameter is adjusted, moving it closer to the optimal value, thereby progressively reducing the overall cost.

Let us look at the gradient descent update rule in more detail via a simple example. Consider a linear regression model with one feature. The model predicts the target value for the i-th observation using the equation:

$$\hat{y}^{(i)} = w_0 + w_1 x_i,$$

where w_0 is the intercept (or bias term) and w_1 is the slope, representing the weight associated with the feature x_i. The error for each observation, also

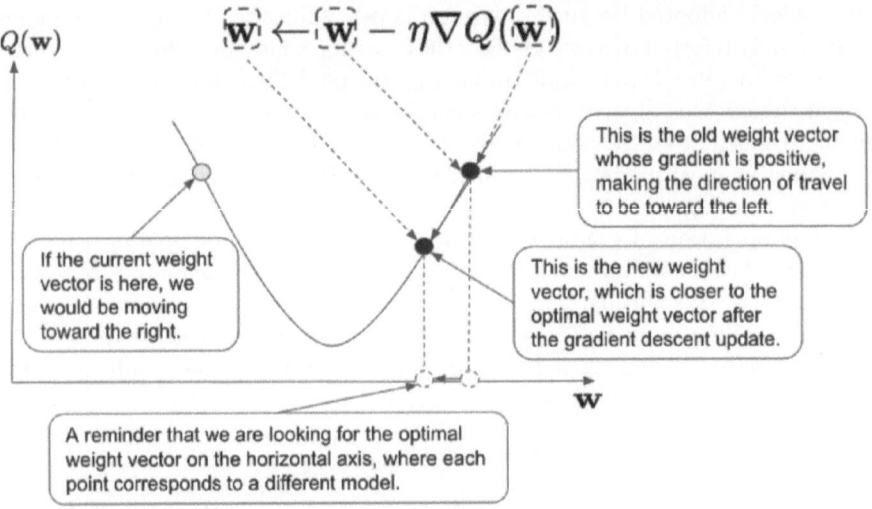

FIGURE 3.6
Gradient descent update rule.

known as the residual, is given by:

$$r^{(i)} = y_i - \hat{y}^{(i)} = y_i - (w_0 + w_1 x_i).$$

The performance of the model is quantified by the SSE cost function:

$$Q(\mathbf{w}) = \sum_{i=1}^{n} (y_i - w_0 - w_1 x_i)^2 .$$

Our objective is to determine the optimal weights w_0^* and w_1^* that minimize $Q(\mathbf{w})$.

To achieve this, we can use gradient descent, which requires us to first compute the gradients of the cost function with respect to each weight. The partial derivative of the cost function with respect to w_0 is:

$$\frac{\partial Q}{\partial w_0} = -2 \sum_{i=1}^{n} (y_i - w_0 - w_1 x_i) ,$$

and the partial derivative with respect to w_1 is:

$$\frac{\partial Q}{\partial w_1} = -2 \sum_{i=1}^{n} x_i (y_i - w_0 - w_1 x_i) .$$

These gradient expressions provide insight into how small changes in each weight affect the cost function. The negative signs indicate that moving in the opposite direction of the gradient will result in a decrease in the cost.

Using these gradients, the update rules for the weights are formulated as:

$$w_0 \leftarrow w_0 - \eta \left(\frac{\partial Q}{\partial w_0} \right) = w_0 + 2\eta \sum_{i=1}^{n} (y_i - w_0 - w_1 x_i),$$

$$w_1 \leftarrow w_1 - \eta \left(\frac{\partial Q}{\partial w_1} \right) = w_1 + 2\eta \sum_{i=1}^{n} x_i (y_i - w_0 - w_1 x_i),$$

where η is the learning rate—a small positive scalar that determines the size of each update step. For practical computational purposes, the constant factor 2 can be absorbed into the learning rate, thereby simplifying the update expressions.

At a higher level, gradient descent treats each set of weights $\mathbf{w} = (w_0, w_1)$ as a point in a two-dimensional parameter space. Associated with each point is a cost $Q(\mathbf{w})$. The algorithm then iteratively adjusts these weights, moving them in the direction that leads to lower cost values, with the ultimate goal of converging to the global minimum of the cost function. In the context of univariate linear regression, this relationship can be visualized as a three-dimensional landscape: the x-axis represents the intercept w_0, the y-axis represents the slope w_1, and the z-axis corresponds to the cost $Q(\mathbf{w})$. Different initial starting points on this "mountain" illustrate how gradient descent navigates the cost surface, ultimately settling at the lowest point, which represents the optimal set of weights.

In the following code listing, we first create arrays of candidate values for both w_0 and w_1 (the intercept and slope, respectively). These candidate values are generated around the optimal weights w^* obtained from a previous computation, and they allow us to evaluate the cost function Q for each unique combination of parameters. In addition, we calculate the partial derivatives (gradients) of the cost function at an initial starting point. This detailed evaluation helps us understand the behavior of the cost function in the parameter space.

```
# Create an array of 100 linearly spaced values around the
    optimal intercept value
w0_vals = np.linspace(w_star[0] - 3, w_star[0] + 3, 100)

# Create an array of 100 linearly spaced values around the
    optimal slope value
w1_vals = np.linspace(w_star[1] - 1, w_star[1] + 1, 100)

# Generate a grid of values to evaluate the cost function across
    the parameter space
w0_grid, w1_grid = np.meshgrid(w0_vals, w1_vals)

# Initialize a grid to store the computed cost values for each (
    w0, w1) pair
Q_grid = np.zeros((100, 100))
for i in range(Q_grid.shape[0]):
    for j in range(Q_grid.shape[1]):
```

```
14      Q_grid[i, j] = ((y - w0_grid[i, j] - w1_grid[i, j] * X)
            **2).sum()
15
16  # Compute the partial derivative of the cost function with
        respect to w0 (intercept)
17  w0_grad = -2 * (y - w[1] * X - w[0]).sum()
18
19  # Compute the partial derivative of the cost function with
        respect to w1 (slope)
20  w1_grad = -2 * (X * (y - w[1] * X - w[0])).sum()
```

Listing 3.9
Evaluating the Cost Function and Gradients for Candidate Parameters.

In the code above, we generate a set of linearly separated candidate values for w_0 and w_1 by using the NumPy function `np.linspace()`. These arrays are then combined into a grid with `np.meshgrid()`, allowing us to compute the cost function Q at each point in the parameter space. The cost at each grid point is computed by summing the squared differences between the actual target values y and the predicted values, based on the candidate parameters. Additionally, we compute the partial derivatives for both w_0 and w_1 at the current starting point, which will guide the update direction in Gradient Descent.

We then visualize the cost function using a contour plot. In this plot, each dashed line represents a level curve of equal cost, providing a clear depiction of the cost landscape with respect to the two parameters. The initial starting point is marked with a star, and an arrow indicates the direction of the next update, which corresponds to the steepest descent path toward a local (or global) minimum. Figure 3.7 (not shown here) displays this contour plot, where the dashed lines represent equal-cost contours, the star marks the initial weights, and the arrow indicates the direction of the first update step.

To summarize, the overall algorithm follows these steps to implement the gradient descent:

1. Initialize Weights: Start with random values for w_0 and w_1.

2. Compute Predictions: Calculate the predicted values $\hat{y}^{(i)}$ using the current weights.

3. Calculate Residuals: Determine the differences between the actual target values and the predictions.

4. Compute Gradients: Calculate the partial derivatives of the cost function with respect to each weight.

5. Update Weights: Adjust the weights by moving them in the direction opposite to the gradients (i.e., the direction of the steepest descent).

6. Iterate: Repeat the process until a convergence criterion is met, such as a minimal change in the cost function.

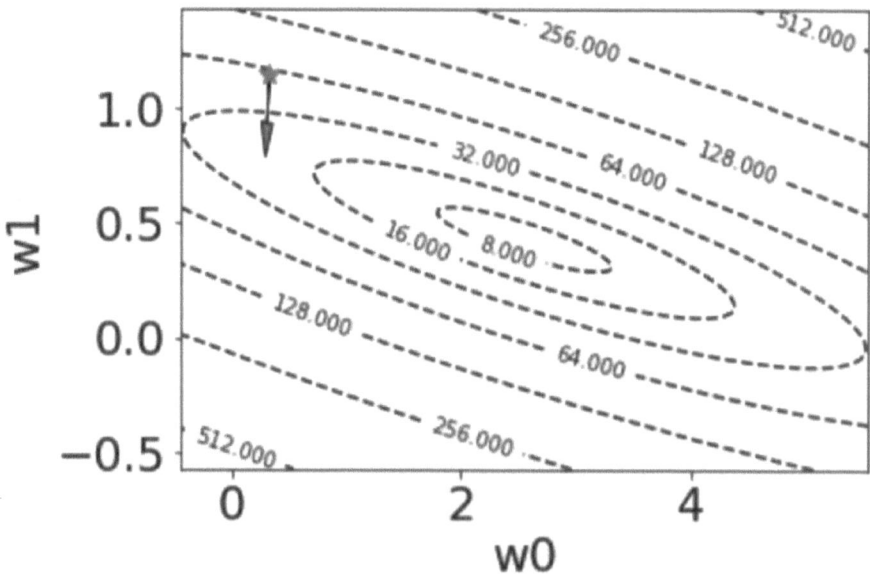

FIGURE 3.7
Contour plot of the cost as a function of two parameters.

This approach of updating the weights by following the negative gradient ensures that the algorithm moves toward regions of lower cost, analogous to descending a valley in a mountainous landscape. The contour plot provides a visual representation of this process, where the weights gradually move toward the minimum cost region with each update.

The following code performs Gradient Descent updates and visualizes the parameter updates at regular intervals.

```
import numpy as np
import matplotlib.pyplot as plt

# Initialize weights randomly with a fixed seed for
    reproducibility
np.random.seed(42)
w = np.random.randn(2, 1)   # [w0, w1]

# Define hyperparameters
learning_rate = 0.01
max_iters = 1000

# Initialize a list to store the history of cost values
cost_history = []

# Setup the plot for visualization of the cost landscape
fig, ax = plt.subplots()
cp = ax.contour(w0_grid, w1_grid, Q_grid, colors='black',
    linestyles='dashed',
```

```
18                           levels=[0, 0.5, 1, 2, 4, 8, 16, 32, 64, 128, 256,
                                 512])
19   ax.plot(w[0], w[1], 'g*', markersize=10)   # Plot the initial
         weights as a star
20   plt.clabel(cp, inline=True, fontsize=10)
21   plt.xlabel('w0')
22   plt.ylabel('w1')
23
24   for i in range(max_iters):
25       # Compute predictions using the current weights
26       predictions = X_new @ w
27
28       # Calculate the residuals (errors) between the true values
             and predictions
29       resid = y - predictions
30
31       # Compute the gradients of the cost function with respect to
             w0 and w1
32       w0_grad = -2 * np.sum(resid)
33       w1_grad = -2 * np.sum(X * resid)
34
35       # Update the weights by moving in the opposite direction of
             the gradients
36       w[0] = w[0] - learning_rate * w0_grad
37       w[1] = w[1] - learning_rate * w1_grad
38
39       # Compute the current cost and append it to the cost history
40       cost = np.dot(resid.T, resid)
41       cost_history.append(cost[0][0])
42
43       # Plot the current weight values every 100 iterations to
             visualize the update path
44       if i % 100 == 0:
45           ax.plot(w[0], w[1], 'g*', markersize=10)
46
47   plt.show()
```

Listing 3.10
Gradient Descent Updates and Parameter Visualization.

The code above performs gradient descent updates and visualizes the parameter updates at regular intervals. Specifically, it initializes the weights randomly and then iteratively refines them by computing the predictions, calculating the residuals, evaluating the gradients, and updating the weights accordingly. At each iteration, the cost is computed and stored in cost_history. The visualization is achieved by plotting the weight values on a contour map of the cost function; here, dashed contour lines represent loci of equal cost, and stars mark the progression of weight updates. As shown in Figure 3.8, the resulting figure illustrates how the weights converge toward the minimum, with the star representing the history of the weight values along the descent path.

To determine when the algorithm should stop updating the weights, the gradient descent algorithm uses one or more stopping criteria, such as:

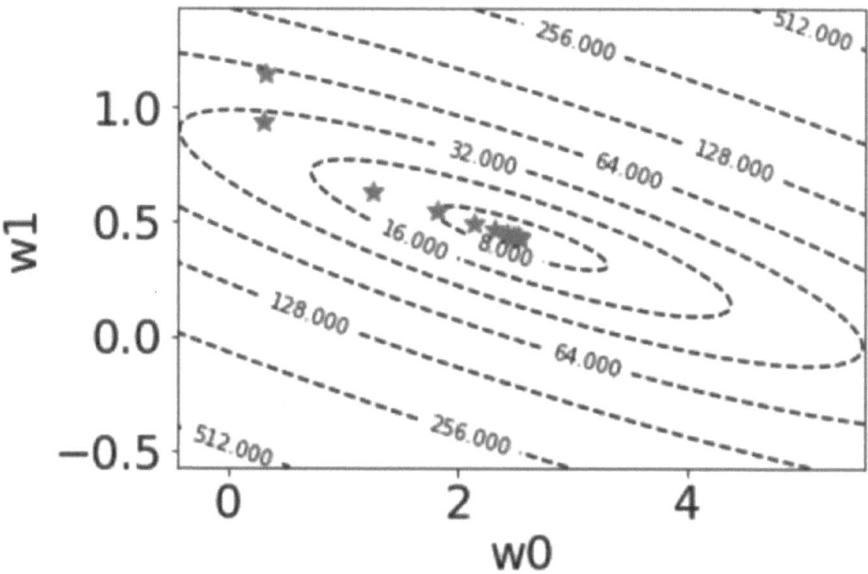

FIGURE 3.8
History of gradient descent updates.

1. Maximum Iterations: The algorithm ceases after a predefined number of iterations have been completed.

2. Convergence Threshold: The algorithm stops when the change in the cost function between consecutive iterations falls below a specified tolerance, indicating that further improvements are negligible.

3. Gradient Norm Threshold: The algorithm terminates when the norm of the gradient vector $\|\nabla Q(\mathbf{w})\|$ is less than a certain threshold, suggesting that the current parameter values are sufficiently close to a stationary point.

Here, introducing a small tolerance parameter can also ensure that the optimization process terminates when additional updates yield minimal improvements, thereby preventing unnecessary computations. For instance, if each update only results in a slight change in the parameters, the algorithm can be stopped early based on an absolute or relative change in the parameter values being smaller than the preset tolerance, even before reaching the maximum number of iterations.

3.5.3 Different types of gradient descent

Gradient descent is a powerful optimization algorithm, but its efficiency can be significantly affected by the size of the dataset. When both the feature matrix

X and the target vector **y** are large, calculating the gradient using the entire dataset at every iteration can become extremely time-consuming and resource-intensive. For example, financial transactional data often comprises millions or even billions of records, making full dataset computations impractical. In such cases, it is useful to consider methods that compute the gradient on only a subset of the data, thereby reducing the computational burden while still steering the model toward a minimum.

In the traditional approach known as Batch Gradient Descent, the gradient is computed by summing the contributions from every training example. This is expressed as:

$$\nabla Q(\mathbf{w}) = \sum_{i=1}^{n} \nabla Q_i(\mathbf{w}) = \sum_{i=1}^{n} \left(-2\mathbf{x}^{(i)}\right)\left(y^{(i)} - \mathbf{w}^\top \mathbf{x}^{(i)}\right),$$

Although this method computes an accurate gradient direction, it becomes prohibitively slow and memory-intensive when n is very large. To address this challenge, alternative strategies have been developed that work with smaller subsets of the data. One such approach is Mini-Batch Gradient Descent, which divides the dataset into smaller groups called mini-batches. Instead of calculating the gradient using the entire dataset or a single data point, the algorithm computes the gradient over each mini-batch. This leads to the update rule:

$$\mathbf{w} \leftarrow \mathbf{w} - \eta \nabla Q_{\mathcal{B}}(\mathbf{w}),$$

where for a mini-batch $\mathcal{B} \subseteq \{1, 2, \ldots, n\}$,

$$\nabla Q_{\mathcal{B}}(\mathbf{w}) = \sum_{i \in \mathcal{B}} \nabla Q_i(\mathbf{w}) = \sum_{i \in \mathcal{B}} \left(-2\mathbf{x}^{(i)}\right)\left(y^{(i)} - \mathbf{w}^\top \mathbf{x}^{(i)}\right).$$

This approach strikes a balance between computational efficiency and convergence stability. By averaging the gradients over a mini-batch, the method reduces the noise inherent in single-sample estimates, leading to smoother updates. Additionally, processing smaller batches requires less memory and allows for parallelization on modern hardware like GPUs (Graphical Processing Units), which can significantly speed up training.

Another widely used variant is stochastic gradient descent (SGD), which updates the weights based on the gradient computed from a single randomly chosen data point at a time. This method is exceptionally light on computational resources since it processes only one example per update. Although the gradient computed from a single example is a noisy approximation of the true gradient, this randomness is not entirely disadvantageous. The inherent noise in the updates can help the algorithm escape local minima, potentially leading to a more robust overall solution. In other words, while the path of the weights during training might appear erratic, the stochastic fluctuations can guide the optimization process toward better generalization (by jumping out of local minima based on stochasticity in the gradient), particularly in the context of very large datasets.

The intuition behind these methods revolves around balancing accuracy with efficiency. Batch gradient descent provides accurate updates by leveraging the entire dataset, but its practicality diminishes as data size increases. SGD, while computationally efficient, may suffer from high variance in its updates. Mini-batch gradient descent offers a compromise by combining the stability of batch methods with the speed of stochastic methods. In practice, the choice among these variants depends on factors such as dataset size, available computational resources, and the specific characteristics of the problem at hand.

3.5.4 The stochastic gradient descent algorithm

SGD is a variation of the Gradient Descent algorithm that updates model parameters using one randomly selected data point at a time. Unlike Batch Gradient Descent, which computes the gradient over the entire dataset during each iteration, SGD updates the parameters incrementally. This incremental approach makes it especially useful in settings like online learning, where data arrives continuously and immediate parameter updates are necessary without processing the complete dataset at every step.

In Batch Gradient Descent, the gradient is calculated by summing the contributions from all data points, ensuring that each update direction is based on the complete picture of the error surface. However, this full-data approach can be computationally expensive when handling large datasets. In contrast, SGD computes the gradient using a single data point, which introduces randomness into the optimization process. Although this randomness means that each update is a noisy approximation of the true gradient, it also allows for more frequent updates and can lead to faster convergence. Additionally, the stochastic nature of SGD can help the algorithm escape local minima, potentially steering the optimization toward a better overall solution.

Recall that Batch Gradient Descent follows the update rule

$$\mathbf{w} \leftarrow \mathbf{w} - \eta \nabla Q(\mathbf{w}) = \mathbf{w} - \eta \sum_{i=1}^{n} \nabla Q_i(\mathbf{w}),$$

where the sum aggregates the gradient contributions from all n data points. In contrast, SGD uses the update rule

$$\mathbf{w} \leftarrow \mathbf{w} - \eta \nabla Q_i(\mathbf{w}),$$

where a single data point i is randomly selected at each iteration. This fundamental difference in computation leads to different convergence behaviors and computational efficiencies. While the updates in SGD are noisier, this noise can be beneficial by encouraging exploration of the error surface and helping the model to avoid getting trapped in local minima.

To illustrate the SGD process, consider a univariate linear regression model. The following example demonstrates how, in one round of SGD, a

single randomly selected data point is used to compute the gradient and update the model parameters:

```
 1   import numpy as np
 2
 3   # Choose a random example
 4   index = np.random.randint(X.shape[0]-1)
 5
 6   # Compute the gradients for the selected example
 7   w0_grad = -2 * (y[index] - w[1] * X[index] - w[0])
 8   w1_grad = -2 * (X[index] * (y[index] - w[1] * X[index] - w[0]))
 9
10   # Update the parameters
11   w[0] = w[0] - learning_rate * w0_grad
12   w[1] = w[1] - learning_rate * w1_grad
```

Listing 3.11
Stochastic Gradient Descent Updates and Parameter Visualization.

In this example, the gradient is computed solely from the selected data
point, and the parameters $w[0]$ and $w[1]$ are updated accordingly. Under SGD,
a complete pass through the dataset—referred to as an epoch—requires n
such individual updates. Typically, training a model involves multiple epochs
to ensure convergence. Although the parameter trajectory under SGD is noisier compared to the smooth path followed by Batch Gradient Descent, this
variability is a key asset. It allows the optimization process to explore the parameter space more widely and can ultimately lead to a more robust solution.

The intuition behind SGD lies in balancing computational efficiency with
effective exploration of the error surface. By making frequent, albeit noisy,
updates, SGD adapts quickly to changes in the data and can adjust the parameters in real-time. This is particularly advantageous in scenarios where
the dataset is massive or when the learning environment is dynamic. In many
practical applications, the benefits of rapid updates and the ability to bypass
suboptimal local minima outweigh the downsides of increased variance in the
parameter updates.

Figure 3.9 shows the parameter traveling path using SGD. Due to the
stochasticity in choosing the example for one step of SGD update, its traveling
path is clearly noisier than the path using full batch gradient descent.

Overall, the stochastic nature of SGD makes it a powerful and flexible optimization tool. Its ability to rapidly update parameters based on individual
data points enables faster convergence in large-scale and streaming data environments, while its inherent randomness provides a mechanism to navigate
complex error landscapes more effectively.

3.5.5 The impact of the learning rate

The learning rate, denoted by η, is a pivotal hyperparameter in the Gradient
Descent family of optimization algorithms. It governs the magnitude of each
step of update taken toward minimizing the cost function, and its careful

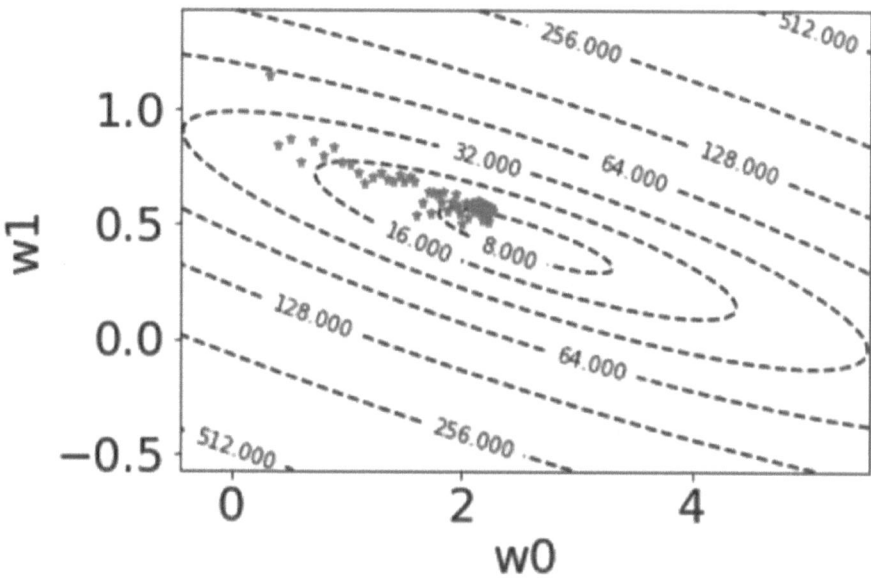

FIGURE 3.9
Parameter traveling path using SGD.

tuning is crucial to ensure efficient and effective convergence of the model parameters. By determining how far the algorithm moves along the negative gradient direction at each iteration, the learning rate directly influences the pace and stability of the optimization process.

The learning rate serves a dual role in the update mechanism. First, it establishes the direction of the update—the negative sign in the gradient descent rule guarantees that adjustments move the weights in a direction that reduces the cost function. Second, it controls the magnitude of these updates, effectively scaling the computed gradient. This scaling is essential: if the updates are too small, progress toward the minimum would be extremely slow, while overly large updates can cause the algorithm to overshoot the minimum, resulting in divergence or oscillatory behavior.

The choice of η thus significantly influences the overall performance of the optimization. A learning rate that is too small results in very small changes to the model parameters, causing the algorithm to require a large number of iterations to approach the optimum. This slow convergence not only extends training times but also increases computational costs, as more iterations are needed to reduce the cost function to an acceptable level. Conversely, a learning rate that is too large can lead to aggressive updates, where the model overshoots the minimum and may even increase the cost function. In some cases, such large steps cause the parameters to oscillate around the minimum without ever converging, making it difficult for the model to stabilize. The intuition behind these dynamics is rooted in a delicate balance between

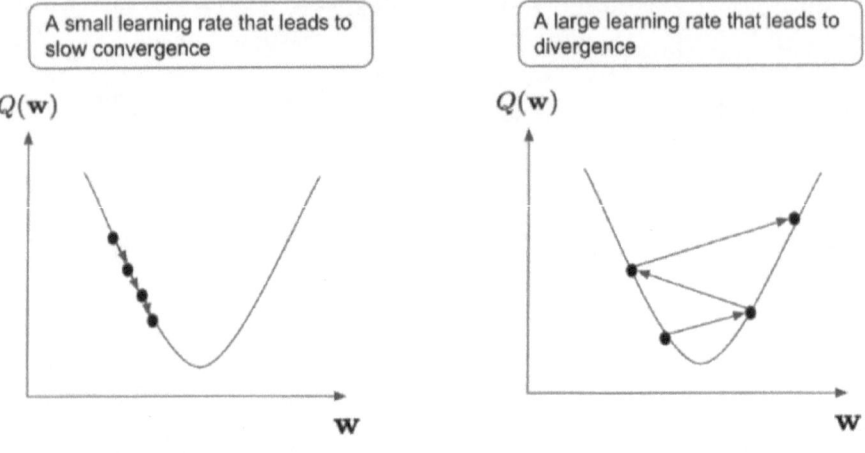

FIGURE 3.10
Two possible scenarios with different learning rates.

precision and progress. The key is finding a balance where the learning rate is sufficiently high to promote rapid convergence without compromising the stability needed to accurately settle at the global minimum.

Figure 3.10 illustrates these two scenarios: the left panel shows how a small learning rate leads to gradual and steady convergence, while another shows that a large learning rate can result in divergence or erratic parameter updates. These visual representations reinforce the idea that the learning rate not only scales the update magnitude but also plays an integral role in determining the overall trajectory of the optimization process.

To evaluate whether the learning rate is appropriately set, one effective method is to plot the cost function $Q(\mathbf{w})$ against the number of iterations. This approach reveals the algorithm's convergence behavior, making it clear whether the chosen learning rate is guiding the model steadily toward the minimum. A steady, consistent decrease in the cost indicates that the learning rate is well-tuned and that the model is making regular progress. Conversely, if the decrease in cost is very slow, it suggests that the learning rate might be too small, leading to slow convergence and prolonged training times. An increasing cost typically signals that the learning rate is too large, causing the updates to overshoot the optimal point and leading to divergence. Similarly, if the cost fluctuates significantly, this indicates oscillations, again a sign of an excessively high learning rate. Figure 3.11 illustrates these scenarios by showing cost curves corresponding to various learning rates, effectively highlighting the different outcomes.

Selecting the right learning rate is a blend of art and science, often requiring experimentation and fine-tuning specific to the problem and dataset at hand. One common strategy is manual tuning: starting with a relatively small value,

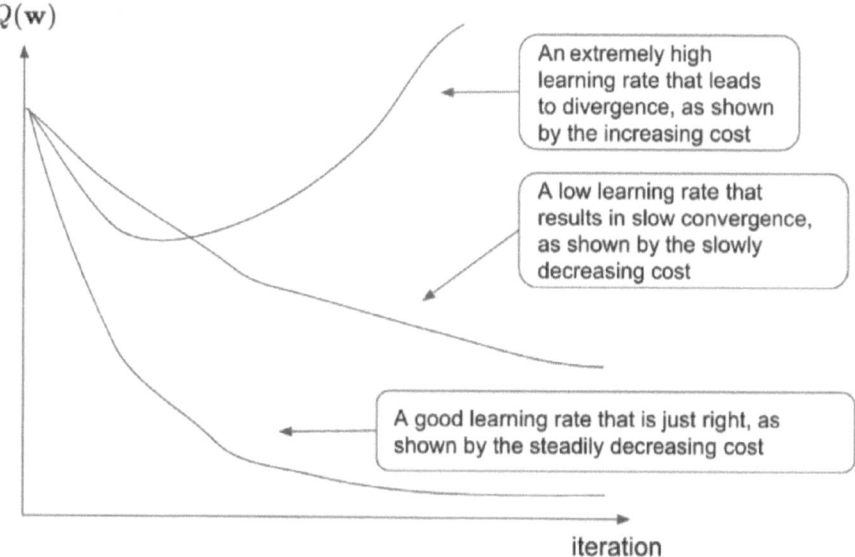

$Q(\mathbf{w})$

An extremely high learning rate that leads to divergence, as shown by the increasing cost

A low learning rate that results in slow convergence, as shown by the slowly decreasing cost

A good learning rate that is just right, as shown by the steadily decreasing cost

iteration

FIGURE 3.11

Plotting the cost as a function of the number of iterations with different learning rates.

such as $\eta = 0.001$, and gradually increasing it while monitoring the cost function. This incremental approach helps identify the point at which the cost begins to diverge, allowing one to adjust the learning rate based on observed convergence behavior.

In addition to manual tuning, learning rate schedules offer systematic approaches for optimization. Fixed schedules maintain a constant learning rate throughout the training process, while decaying schedules gradually reduce the learning rate as training progresses, allowing the algorithm to fine-tune parameters as it nears the minimum. Adaptive schedules go a step further by modifying the learning rate in response to the model's performance during training. In addition, adaptive learning rate algorithms such as Adam, RMSprop, and Adagrad dynamically adjust the learning rate for each parameter using historical gradients. These methods enhance convergence speed and stability by automatically balancing the need for rapid progress with the caution required to avoid overshooting.

3.6 Improving Predictive Performance

The predictive performance of a machine-learning model is intrinsically tied to its complexity, which is often determined by the number of adjustable weights (parameters) within the model. An underfitting model typically has too few weights, limiting its capacity to capture the underlying patterns in the data, while an overfitting model contains an excessive number of weights, enabling it to memorize the training data yet perform poorly on unseen data. To improve predictive performance on the training set, one strategy is to increase the effective number of adjustable weights, thereby enhancing the model's capacity to learn complex relationships. However, this approach must be carefully calibrated using a validation set and subsequently evaluated on a test set to ensure that the model generalizes well rather than simply memorizing the training data.

In the context of a linear regression model, the weight vector $\mathbf{w}$ typically contains two elements corresponding to the intercept and the slope. Such a model predicts the target $\hat{y}^{(i)}$ for the i-th instance using the equation

$$\hat{y}^{(i)} = w_0 + w_1 x^{(i)}$$

where w_0 represents the intercept (or bias term) and w_1 is the weight associated with the feature $x^{(i)}$. This simple structure highlights how each component of the model directly contributes to the prediction, which is one of the reasons linear models are so valued for their interpretability.

Although linear models are favored for their explainability (since any change in the prediction is directly and linearly related to changes in the input feature), they often struggle to capture the nonlinear patterns that are present in real-world data. To address this limitation while still retaining the good interpretability of linear models, feature engineering can be used. By generating additional features based on the original ones, feature engineering expands the model's capacity to fit more complex and nonlinear relationships without compromising its inherently linear nature. This process allows practitioners to transform or combine existing features in ways that reveal underlying patterns that a simple linear model might otherwise miss.

One widely used technique to enlarge the feature space of linear regression model is to create additional polynomial features based on existing ones.

3.6.1 Polynomial feature engineering

A common form of feature engineering is the use of polynomial basis functions. By introducing polynomial terms, we can transform the original one-dimensional feature space into a higher-dimensional space, enabling the model to capture curvatures and nonlinear trends in the data. For example, enhancing the feature $x^{(i)}$ for the instance i^{th} with its square $(x^{(i)})^2$ allows the model to fit a quadratic curve:

$$\hat{y}^{(i)} = w_0 + w_1 x^{(i)} + w_2 (x^{(i)})^2$$

Here, the weight vector $\mathbf{w} = [w_0, w_1, w_2]^\top$ now contains three parameters, offering greater flexibility in modeling the data by accounting for curvature and

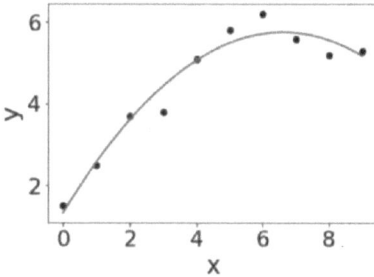

FIGURE 3.12
A nonlinear curve fitting using polynomials up to the second degree.

more subtle variations. The quadratic transformation augments the original feature space by adding a squared term, resulting in a new design matrix:

$$\Phi(\mathbf{X}) = \begin{bmatrix} 1 & x^{(1)} & (x^{(1)})^2 \\ 1 & x^{(2)} & (x^{(2)})^2 \\ \vdots & \vdots & \vdots \\ 1 & x^{(n)} & (x^{(n)})^2 \end{bmatrix}$$

This transformation not only expands the feature space but also enhances the model's capacity to capture more complex relationships within the data. The prediction function becomes:

$$\hat{\mathbf{y}} = \Phi(\mathbf{X})\mathbf{w} = w_0 + w_1 x^{(i)} + w_2 (x^{(i)})^2$$

By introducing polynomial features, the model is able to fit nonlinear trends into the data. This increased flexibility can result in a lower SSE, as the model can more accurately represent the underlying data patterns by adapting to the curvature in the data. However, it is essential to balance this flexibility to avoid overfitting, where the model becomes excessively tailored to the training data and performs poorly on new, unseen data. The intuition here is that while adding polynomial features enriches the model's representational power, it also increases the risk of fitting noise; therefore, careful validation and regularization are often exercised to maintain generalizability.

Figure 3.12 demonstrates how to augment the feature space using polynomial features up to the second degree. It shows that the nonlinear trend in the data has been successfully captured by the new quadratic model due to the augmented features up to the second degree of polynomials. Here, a nonlinear curve is fit using polynomials up to the second degree. The resulting cost is significantly reduced to approximately 0.94, compared to the previous cost of 6.3 based on the original linear model up to the first degree, underscoring the efficacy of polynomial feature engineering in enhancing the model's predictive performance by providing a better fit to the true underlying data structure.

3.6.2   Linearity in the weights

It is important to note that the nonlinear transformation introduced by polynomial feature engineering is applied solely to the input design matrix $\Phi(\mathbf{X})$ and not to the weight vector $\mathbf{w}$. As a result, the model still remains linear in its weights:

$$\hat{\mathbf{y}} = \Phi(\mathbf{X})\mathbf{w}$$

This formulation implies that although the relationship between the input features and the target variable may be rendered nonlinear through the feature transformation, the model is still considered as linear because it depends linearly on the parameters $\mathbf{w}$. This inherent linearity not only preserves the interpretability of the model but also facilitates a clear and straightforward analysis of how each parameter contributes to the final predictions, allowing for both efficient optimization and transparent inference of the model's behavior.

In contrast, nonlinear models such as neural networks introduce nonlinear transformations that extend to the weights themselves, thereby making the model nonlinear with respect to its parameters and consequently more complex to interpret. Although these models possess the capability to capture intricate and highly nonlinear patterns within the data, they often sacrifice the direct explainability characteristic of linear models. This complexity can complicate the process of understanding the precise influence of individual weights, thus making it more difficult to analyze how specific parameters drive the predictions.

3.7   More on the Model

We have been using $f_{\mathbf{w}}$ to represent the model f developed based on a set of weights $\mathbf{w}$, but for a broader discussion, we will denote the model simply as f. It is worth noting that model choice can vary along several dimensions, such as linearity, where models may be linear or nonlinear, and the nature of parameterization, with parametric models having a fixed number of parameters and nonparametric models allowing the number of parameters to increase with the amount of data available. For example, in linear regression, the model f is inherently linear and parameterized by $\mathbf{w}$, a formulation that not only clarifies its structure but also aids in understanding the underlying assumptions.

Model training involves building a model that represents the true mapping relationship between the input variable x and the output variable y. Let us denote the underlying real mapping function as g, which is nonrandom, consistent, and encapsulates the true relationship between x and y. Our objective is to estimate this function g as accurately as possible using our model

f, thereby ensuring that the predictions produced by f closely reflect the true dynamics inherent in the data-generating process.

Assuming an additive random noise ϵ in the data-generating process, we can express the underlying relationship as

$$y = g(x) + \epsilon.$$

In this expression, ϵ is an additive noise term that is assumed to follow a normal distribution with zero mean and fixed variance σ^2, that is,

$$\epsilon \sim \mathcal{N}(0, \sigma^2).$$

This random variable ϵ accounts for errors arising from measurement inaccuracies or other real-world sources of noise and is assumed to be independent of the input variable x in the regression setting. Therefore, the goal of linear regression is to construct a linear model f that approximates the true function g as closely as possible, even in the presence of this additive, independent, and random noise.

Given the limited representativeness of the available dataset $(\mathbf{X}, \mathbf{y})$ for capturing all possible variations from the distribution $P(x, y)$, we often aim to build a model f whose expected output matches that of g. Mathematically, we choose a model f such that

$$\mathbb{E}[f(X)] \approx \mathbb{E}[y \mid x = X] = \mathbb{E}[g(x) + \epsilon \mid x = X] = \mathbb{E}[g(X)] + \mathbb{E}[\epsilon] = g(X).$$

This equation ensures that, on average, our model f aligns with the true mapping function g, thereby confirming that the model is capable of generalizing appropriately despite the inherent limitations of any finite dataset.

3.7.1 Bias and variance decomposition

As introduced in Chapter 1, when training a model f, there are three primary sources of error to balance: bias, variance, and the irreducible error due to inherent noise in the data. Bias refers to the error due to the model's insufficient capacity to perfectly fit the true function g, and it represents the underfitting scenario. For instance, when a model is too simplistic relative to the underlying complexity of g, such as fitting a nonlinear quadratic pattern using a linear model f, the result is high bias. In contrast, variance refers to the error arising from the model capturing spurious relationships that are corrupted by the random noise ϵ present in the data, thereby representing the overfitting scenario. For example, even when fitting a linear pattern g with a linear model f, if the model fails to capture the true underlying structure (i.e., $\mathbb{E}[f] \neq g$), it can exhibit high variance by overly adapting to the noise in the training data.

Following the discussion on the estimation risk in Chapter 2, we can decompose the expected risk for a new, unseen input X_{new}. For notational simplicity, we denote $f(X_{\text{new}}) = f$ and $g(X_{\text{new}}) = g$. Through careful mathematical

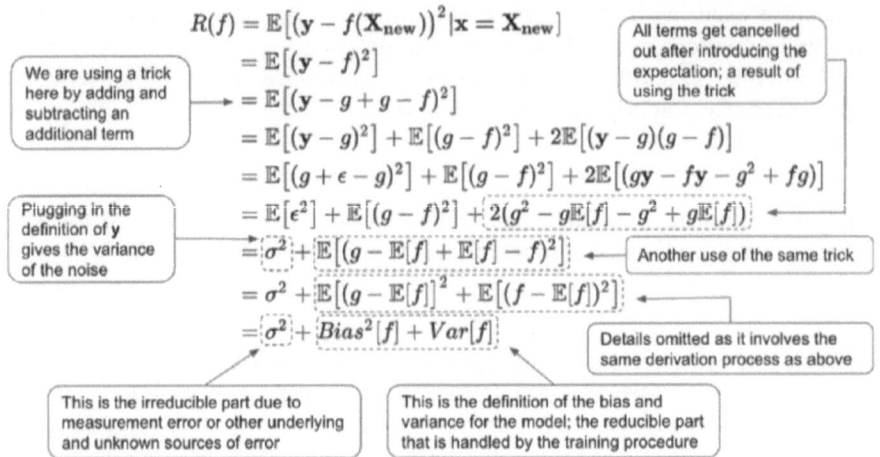

FIGURE 3.13
Decomposing the risk into the irreducible part, that is, the variance of the error term, and the reducible part, that is, the squared bias and variance of the model.

derivation, the expected squared error can be decomposed into three distinct components: the squared bias, $(\text{Bias}[f])^2$; the variance of the model, $\text{Var}[f]$; and the variance of the noise, σ^2 (which represents the irreducible error). This decomposition can be summarized by the following:

$$\mathbb{E}[(y - f)^2] = (\text{Bias}[f])^2 + \text{Var}[f] + \sigma^2,$$

which not only clarifies the individual contributions of each error source but also highlights the inherent limitations in reducing the overall error.

Figure 3.13 illustrates this decomposition process, highlighting the algebraic trick of introducing an additional term—by adding and subtracting the same quantity—when expanding the squared error expression. This technique simplifies the derivation and makes the underlying structure of the decomposition clearer.

This process effectively separates the risk into two distinct components: the irreducible part, represented by the variance of the error term, and the reducible part, which comprises both the squared bias and the variance of the model. This clear separation aids in understanding which aspects of the error can be mitigated through improved modeling techniques.

The above derivation also uses the facts that $\mathbb{E}[g] = g$ (since g is deterministic and nonrandom) and $\mathbb{E}[\epsilon] = 0$, where y is the observed target for X_{new}. Additionally, it follows that

$$\mathbb{E}[fy] = \mathbb{E}[f]\mathbb{E}[y] = \mathbb{E}[f]\mathbb{E}[g + \epsilon] = \mathbb{E}[f](\mathbb{E}[g] + \mathbb{E}[\epsilon]) = g\,\mathbb{E}[f].$$

These simplifications rely on the fundamental properties of expectation for deterministic functions and independent random noise, ensuring that the expected value of the true function remains unchanged and that the noise term does not bias the mean error.

Minimizing the risk involves reducing one or more of these three sources of error. The random noise term, quantified by σ^2, is an irreducible source of error that cannot be controlled, while both bias and variance are reducible elements that can be managed and optimized through the training procedure. In practice, efforts to minimize overall risk focus on decreasing the squared bias and the model's variance, acknowledging that the noise component remains an inherent aspect of the data-generating process.

As discussed earlier, a simple model tends to incur a high bias and low variance, leading to underfitting, whereas a complex model typically exhibits low bias and high variance, resulting in overfitting. This observation aligns with the classical statistical view on the bias-variance tradeoff, where reducing one often leads to an increase in the other. The interplay between bias and variance imposes a fundamental limitation on a model's capability to generalize to future test data, as it is challenging to simultaneously minimize both sources of error. Together, these three sources of error establish a lower bound on the expected error for any future unseen dataset.

However, as will be explored in later chapters, over-parameterized models such as smoothing splines and neural networks can continue lowering the variance even when the bias has already been reduced to zero by exploring a better generalizing solution in the parameter space. This phenomenon, known as double-descent, challenges traditional conceptions of the bias-variance tradeoff and provides new insights into the behavior of complex models in high-dimensional settings.

3.7.2 Understanding bias and variance using bootstrap

Recall that the model bias refers to the gap between the expected model prediction $\mathbb{E}[f]$ and the underlying true target g:

$$\text{Bias}[f] = g - \mathbb{E}[f]$$

The expectation $\mathbb{E}[f]$ considers all different realizations of the dataset $(\mathbf{X}, \mathbf{y})$ sampled from the same joint data-generating distribution $P(x, y)$, effectively averaging the model's behavior over multiple training scenarios. A biased model is said to underfit the training data, as it fails to capture the true relationship between the input features and the output label, resulting in predictions that consistently deviate from the true target. A less biased model can better capture the regularities in the data and represent the underlying complexity more effectively, leading to improved accuracy and more reliable inference.

In the previous decomposition formula, the squared bias term can be interpreted as the error introduced by a simplifying assumption in the model, such as approximating an inherently nonlinear pattern g using an under-capacitated linear model f. This assumption leads to an insufficient representation of the true pattern in the data, which in turn contributes to systematic prediction errors. Increasing the model complexity by using feature augmentation or incorporating additional useful parameters is likely to enhance the model's fitting capacity and thus reduce its bias by providing a better approximation of g. However, this increase in complexity may also risk overfitting the data if the model begins to capture noise as if it were a signal.

The variance measures the expected movement of f around its mean:

$$\mathrm{Var}[f] = \mathbb{E}[(f - \mathbb{E}[f])^2]$$

This quantity refers to the model's sensitivity to possible changes in the data and reflects how much the prediction f can fluctuate when different training samples are used. As a result of overfitting the random noise, a model with high variance becomes overly complex and sensitive to even small fluctuations in the data. Note that the level of complexity for a high-variance model f is defined in terms of the current dataset and may not accurately represent the true complexity of the underlying function g, which remains unknown in practice.

Since both bias and variance are highly dependent on the available dataset $(\mathbf{X}, \mathbf{y})$, one effective method to quantify a model's bias and variance across different datasets is the bootstrap procedure used to simulate additional artificial datasets. Bootstrap is a technique to generate artificial datasets by randomly sampling with replacement from the original dataset $(\mathbf{X}, \mathbf{y})$, thereby creating multiple pseudo-replicates that reflect the variability inherent in the data. These artificial datasets serve as additional realizations from $P(x, y)$ and can be used to perform the same model-fitting procedure multiple times. The resulting multiple fits can then be used to estimate the bias and variance of the original model f, providing valuable insight into its stability and generalization performance.

Figure 3.14 provides a visual illustration of the estimated bias and variance using bootstrap. The leftmost plot shows the underlying function and actual observations disrupted by random noise, offering a clear view of the true data-generating process. The middle plot displays the bias by highlighting the difference between the fitted model and the true function, while the rightmost plot presents the multiple fits based on bootstrapped samples as dashed lines, with the original fitted model shown as a solid line. The spread among these dashed lines visually demonstrates how sensitive the model is to the bootstrapped training data and showcases the variability of the model.

Visualizing the three components of the model's risk after introducing bootstrap offers a concrete way to understand how bias, variance, and noise contribute to the overall error. The bootstrap procedure is a powerful technique for generating artificial datasets and is also connected to the concept of assembling, where multiple fits are performed and aggregated to produce the final prediction. The additional fits can stem from new datasets or even different model architectures, each contributing a regularizing effect either via

$$R(f) = \sigma^2 + Bias^2[f] + Var[f]$$

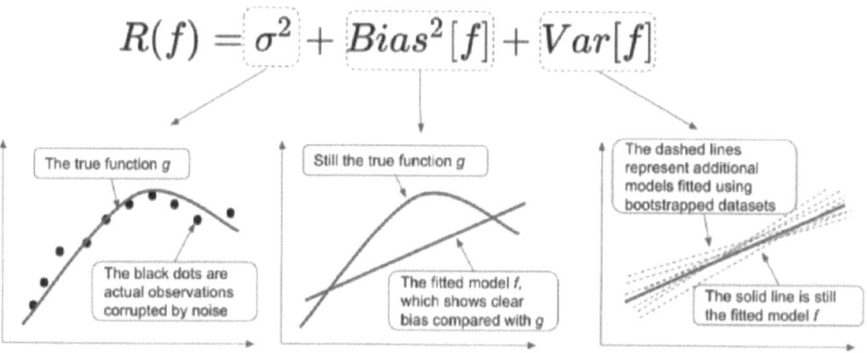

FIGURE 3.14
Visualizing the three components of the model's risk after introducing bootstrap.

the data or the model itself and ultimately leading to more robust predictive performance.

3.7.3 Reduced generalization with high model complexity

The classical view on generalization posits that the two reducible sources of error, namely bias and variance, cannot be reduced simultaneously. When the bias is high and the variance is low, the model is relatively simple and tends to underfit the training data, a situation that often arises at the initial stage of training. In this scenario, a common strategy for empirical risk minimization is to train a more complex model. Increasing the model complexity typically leads to a relatively large decrease in bias while incurring only a slight increase in variance, thereby reducing the overall combined error. Although this increased variance implies that the model's predictions may become more sensitive to fluctuations in the training data, potentially resulting in outputs that deviate from the true target when new data are encountered, the general benefit of reducing bias by increasing model complexity often outweighs the downside of a somewhat less stable estimate when the goal is to build a generalizable model.

The classical view also indicates that a high-variance model is less likely to generalize well to future test sets. When a model becomes arbitrarily complex, it tends to overfit the training data and may even perfectly interpolate it, resulting in zero training error. However, such a model often suffers from high variance because it fits not only the underlying signal but also the noise present in the training data. Consequently, when using the model to score new data points, even a slight change in these points can lead to very different outputs, making the predictions both sensitive and unstable. In contrast, a well-balanced model achieves an appropriate tradeoff between bias and variance, thereby avoiding the pitfalls of underfitting and overfitting. This delicate balance, often described as the "sweet spot" in Chapter 1, underscores the importance of carefully tuning model complexity to achieve robust generalization performance.

3.7.4   Observing increased variance by varying model complexity

To observe the increased variance as the model becomes complex, we employ additional feature engineering by increasing the number of polynomial basis functions to a higher degree. Instead of running the same codes multiple times, each with a different polynomial degree, it is more efficient to encapsulate the polynomial transformation into a general function. Designing a reusable function enhances flexibility in specifying the degree of the polynomial transformation and streamlines the experimentation process. The following function achieves this by creating a matrix as a placeholder based on the predefined degree of the polynomial, followed by filling in the matrix using the appropriate polynomial functions.

```
def polynomial(x, num_basis=2):
    # Create the matrix of zeros as a placeholder
    Phi = np.zeros((x.shape[0], num_basis))
    # Fill in each column based on the sequence of polynomials
    for i in range(num_basis):
        Phi[:, i:i+1] = x**i
    return Phi
```

Listing 3.12
Creating Polynomial Features.

We can then create a new design matrix using this utility function by passing in the original design matrix and specifying the number of basis functions required. Figure 3.15 shows the fitted model using three different numbers of basis functions along with the respective cost associated with each configuration. This figure clearly illustrates that the model becomes increasingly unstable as the number of basis functions grows, indicating a rise in variance. Specifically, the left plot shows a better fit compared to the model using up to the second-degree polynomial. When the degree of the polynomial grows to 10 or even 15, the model perfectly fits all the data points, effectively capturing every nuance of the training set. However, this comes at the cost of high variance, especially in the middle and right regions of the plot, where the predictions become highly erratic due to the model overfitting to the noise. For further details and the full training codes, please refer to the accompanying notebook.

The figure above also illustrates that the fitted model is less generalizable to the test set when high variance is present as a result of excessive model complexity. The bias-variance tradeoff is thus central to classical generalization theory: the fitted model needs to find the "sweet spot" and

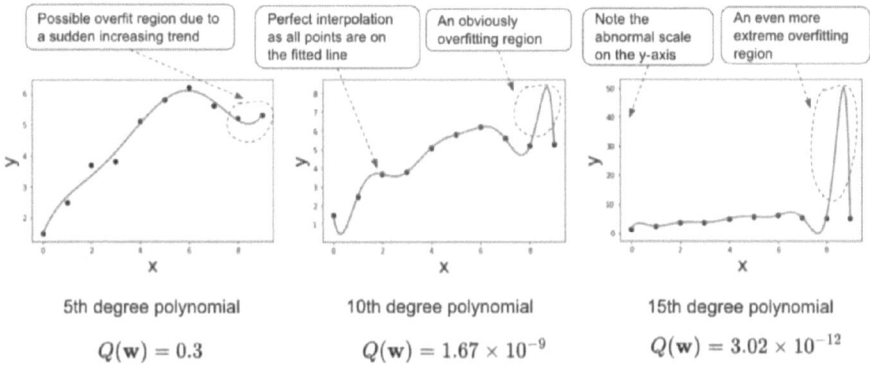

FIGURE 3.15
Fitting three models using different numbers of polynomial basis functions.

properly balance underfitting and overfitting. A good model should be expressive enough to represent the underlying pattern in the data while remaining simple enough to avoid fitting potentially spurious patterns caused by random noise, ensuring that it performs robustly on new, unseen data.

3.8 Summary

In this chapter, we went through an exploration of classical generalization from the perspective of linear regression, laying the groundwork for understanding more complex modeling paradigms later on. We began by reviewing the essential elements of supervised learning, where the primary objective is to learn a mapping function $f : \mathbb{R}^d \to \mathbb{R}$ that approximates the true underlying function g in the presence of additive noise. The discussion detailed how the inductive bias inherent in models such as linear regression—where the mapping is assumed to be linear—shapes both the formulation and the subsequent learning process.

The training pipeline was decomposed into four fundamental components: data, model, cost function, and optimization. We examined the structure of the dataset, emphasizing that data are typically assumed to be i.i.d. samples drawn from an unknown distribution $\mathcal{P}(\mathbf{x}, y)$. The train-test split was discussed as a critical step to mitigate overfitting and ensure that the model performance is evaluated in a manner representative of its generalization capability.

For the working mechanism of linear regression, we described the model as a linear mapping $\hat{\mathbf{y}} = \mathbf{\Phi}\mathbf{w}$ where the design matrix $\mathbf{\Phi}$ may be augmented via the bias trick to incorporate an intercept or feature engineering, and the weight vector $\mathbf{w}$ encapsulates the model parameters. The cost function, taking the form of SSE or MSE, was shown to possess a quadratic structure that permits a closed-form solution via the Normal Equation $\mathbf{w}^* = \left(\mathbf{\Phi}^\top \mathbf{\Phi}\right)^{-1} \mathbf{\Phi}^\top \mathbf{y}$. This convexity guarantees the uniqueness of the global minimum, a key advantage in the linear regression context.

We then explored iterative optimization methods, primarily focusing on gradient descent and its variants (mini-batch and SGD). The gradient descent algorithm was analyzed both conceptually and through detailed derivation, highlighting the importance of the learning rate η in balancing convergence speed against the risk of overshooting the minimum. We used contour plots and parameter trajectories to illustrate how the algorithm navigates the cost landscape.

Recognizing that predictive performance hinges on the model's ability to capture the underlying data structure without overfitting, we introduced strategies for enhancing model capacity via polynomial feature engineering. By augmenting the original feature space with polynomial terms, the linear model can represent nonlinear relationships while retaining linearity in the weights. This transformation increases the expressiveness of the model and reduces bias; however, it also introduces the risk of high variance if not carefully regularized.

The classical bias-variance tradeoff was then revisited. We decomposed the expected prediction error as

$$\mathbb{E}[(y - f)^2] = (\text{Bias}[f])^2 + \text{Var}[f] + \sigma^2,$$

and discussed how underfitting (high bias) and overfitting (high variance) represent two sides of the same coin. Techniques such as bootstrap were presented as practical tools to empirically assess and visualize bias and variance, thus providing guidance for model selection and complexity control.

Finally, we emphasized that the goal of model training is to approximate the true function g while balancing the competing demands of bias reduction and variance control. Achieving this balance is central to the generalization of machine-learning models, ensuring that they perform robustly on unseen data.

In summary, this chapter provided a theoretical and practical foundation for classical generalization. It connected the fundamental principles of supervised learning, linear modeling, and optimization to the broader challenge of balancing model complexity, a balance that lies at the heart of effective predictive modeling.

4

Modern Perspectives on Generalization

One of the most important insights in previous chapters is the bias-variance tradeoff, a fundamental concept in statistical learning theory that specifies the intricate relationship between a model's complexity and its ability to generalize to unseen data. In this chapter, to comprehensively grasp this tradeoff, we will delve deeper into the underlying mathematics and explore how it connects with contemporary observations in machine learning.

As explained in Chapter 3, the expected prediction error for a model can be decomposed into three distinct components:

$$\text{Error} = \text{Bias}^2 + \text{Variance} + \text{Irreducible Error}$$

This decomposition provides a framework for understanding the sources of error in statistical models:

1. Bias (squared): The first component measures the error introduced by approximating a real-world problem, which may be inherently complex, by a simplified (and thus biased) model. It is defined as the squared difference between the expected prediction of the model and the true underlying function:

$$\text{Bias}^2 = \left(\mathbb{E}[\hat{f}(x)] - f(x) \right)^2$$

Here, $\mathbb{E}[\hat{f}(x)]$ represents the expected prediction of the model for a given input x, while $f(x)$ is the true underlying function, which is typically unknown.

2. Variance: This term quantifies the model's sensitivity to fluctuations in the training dataset. It represents the variability of the model prediction for a given data point x across different possible training sets:

$$\text{Variance} = \mathbb{E}\left[\left(\hat{f}(x) - \mathbb{E}[\hat{f}(x)] \right)^2 \right]$$

This equation captures how much the model's predictions would change if it were trained on different subsets of the data. It is essentially the definition of variance for $\hat{f}(x)$ as a random variable.

3. Irreducible Error: This is the noise inherent in the data, which cannot be eliminated regardless of the model used. It represents the fundamental uncertainty in the problem domain.

The bias-variance tradeoff posits that as model complexity increases, bias tends to decrease while variance tends to increase. In contrast, reducing model

DOI: 10.1201/9781003511601-4

complexity tends to increase bias and decrease variance. This relationship creates a delicate balance in model selection and tuning when one seeks the "sweet spot" of model complexity. The optimal model achieves an equilibrium where the sum of bias squared and variance is minimized, thereby minimizing the total prediction error. Understanding this tradeoff is crucial, as it guides the process of model selection and helps in diagnosing issues of overfitting or underfitting in predictive models.

Traditionally, the goal in model selection is to find a model that generalizes well to unseen data. The classical view on generalization, as discussed in Chapter 3, posits that a model with excessive complexity, one that has too many parameters relative to the amount of training data, will incur high variance. This high variance is expected to lead to overfitting, where the model over-tunes itself not only to the underlying true relationships but also to the random noise in the training data. Consequently, while such a model may perform exceptionally well on the training set, its performance is expected to deteriorate significantly on new, unseen test data.

To further analyze this tradeoff, consider a model complexity parameter d (such as the degree of a polynomial in a linear regression model). As d increases, we typically observe:

$$\text{Bias}^2 \downarrow \quad \text{and} \quad \text{Variance} \uparrow$$

According to this classical understanding, the optimal complexity of model d^* is achieved at a "sweet spot" where the test error is minimized, comprising both bias and variance is minimized. Graphically, this is often depicted as the bottom of an inverted U-shaped risk curve. On this curve, the test error initially decreases with increasing d, reaches a minimum at d^*, and then starts to increase as d continues to grow beyond this optimal point.

However, an intriguing phenomenon in contemporary machine-learning practice, particularly with over-parameterized models like deep neural networks, seems to challenge this classical view. These models often have more parameters than training data points, which, according to traditional theory, should lead to high variance and poor generalization. However, despite achieving zero training error, which indicates perfect fit or even apparent overfitting, they frequently demonstrate good generalization to test data.

This scenario can be considered as an over-parameterized interpolating model where the number of parameters p exceeds the number of data points n:

$$p > n \quad \Rightarrow \quad \text{Training Error} = 0$$

Surprisingly, despite the zero training error, these models do not exhibit the expected high variance on the test set. Instead, they often manage to capture the underlying data distribution effectively without succumbing to the noise present in the training data. This phenomenon, sometimes referred to as "benign overfitting," has sparked significant interest and research in the machine-learning community.

The apparent contradiction between classical statistical learning theory and the empirical success of over-parameterized models in modern machine-learning practice has led to new lines of research. Many are now investigating the conditions under which over-parameterization can lead to good generalization and how factors such as model architecture, optimization algorithms, and data properties contribute to this unexpected behavior. In this chapter, we will attempt to provide more insight into this phenomenon.

4.1 A Modern View on Generalization

When training a machine-learning model, our primary objective is to identify the optimal model from a (possibly infinite) set of potential candidate models. This optimal model is characterized by its ability to make the fewest mistakes on a previously unseen test dataset, thus demonstrating superior generalization capabilities.

Formally, we can express this objective as finding the optimal model f_w^* from a set of candidate models $\mathcal{F}_w$ that minimizes the average loss on a training dataset (referred to as the empirical loss):

$$f_w^* = \arg \min_{f_w \in \mathcal{F}_w} \frac{1}{n} \sum_{i=1}^{n} L\big(y^{(i)}, f_w(x^{(i)})\big)$$

In this equation, f_w^* represents the optimal model, and $\mathcal{F}_w$ is the set of all candidate models parameterized by w, $L\big(y^{(i)}, f_w(x^{(i)})\big)$ is the loss function that quantifies the discrepancy between the true label $y^{(i)}$ and the model's prediction $f_w(x^{(i)})$, and n is the number of samples in the dataset. The challenge lies in finding a model that strikes the right balance between learning from the training data and maintaining the ability to generalize to new, unseen data. This balance is crucial in avoiding two common pitfalls: overfitting, where the model becomes too specific to the training data, capturing noise and idiosyncrasies that don't generalize well; underfitting, where the model is too simplistic and fails to capture important patterns in the data.

The goal is thus to develop a model that not only performs well on the training data but also maintains its performance when dealing with new, unseen examples. This ability to generalize is the hallmark of a truly effective generalizable machine-learning model, allowing it to make accurate predictions or decisions in real-world scenarios beyond its training environment.

The classical view on generalization in machine learning is characterized by a distinctive U-shaped curve that illustrates the bias-variance tradeoff. This curve typically represents the model's performance on the test set, measured as the empirical test risk, and plotted as a function of model complexity. To provide a comprehensive understanding, the training set's performance curve

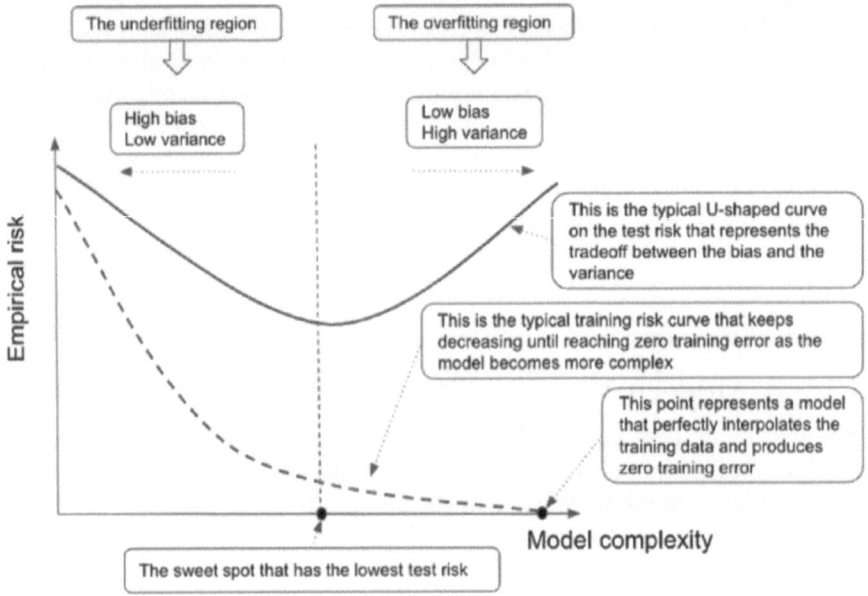

FIGURE 4.1
The typical U-shaped training and test risk curves that indicate the classical view on the bias-variance tradeoff.

is often plotted alongside it, demonstrating the model's increasing predictive power on the training data.

The U-shaped curve, as illustrated in Figure 4.1, reveals several key insights. Starting with the initial decrease, as the model complexity increases from a very simple state, both training and test risks decrease. This indicates that the model is capturing more of the underlying patterns in the data. Coming to the divergence point, at a certain level of complexity, the training risk continues to decrease, but the test risk begins to increase. This divergence marks the onset of overfitting. As the model becomes overly complex, it overfits the training, data and the divergence continues. Consequently, while the training risk continues to decrease, potentially reaching zero, the test risk rises, indicating poor generalization.

According to this classical perspective, the optimal model is found at the "sweet spot," which is the point of minimum empirical test risk. This point represents the ideal balance between underfitting and overfitting. Specifically, the left of the sweet spot represents an underfitting regime where the model is too simplistic and fails to capture the true relationships in the data. In this case, the (squared) bias term dominates the test risk, indicating that the model's predictions are consistently off-target. The right side of the sweet spot represents an overfitting regime where the model has become overly complex, fitting noise in the training data rather than the underlying pattern. Here, the variance term dominates the test risk, resulting in unstable predictions that don't generalize well to new data.

This classical view provides a framework for model selection and complexity tuning. It suggests that the goal in model development should be to find the right level of complexity that minimizes test risk (often approximated via an additional validation set), thereby achieving the best balance between bias and variance. However, it is important to note that recent developments in deep learning have challenged some aspects of this classical view, particularly in cases of highly overparameterized models. These developments have led to ongoing research and debates about the nature of generalization in modern machine-learning systems.

Holding the classical view on model complexity and generalization, we have seen different mechanisms to control model performance through regularization, including data manipulation, model architecture design, cost function selection, and optimization algorithm choice. According to this perspective, when model complexity is carefully balanced to avoid both underfitting and overfitting, the model is expected to generalize well to unseen data drawn from the same population as the training set. This understanding has led to the widespread use of the U-shaped risk curve as a guide for model selection.

While classical statistics suggest that larger, more complex models would suffer in predictive performance once they overfit the training data, recent observations in machine learning have challenged this view. Practitioners often find that certain ML models continue to achieve lower test errors even as they become increasingly complex, far beyond the point where classical theory would predict overfitting. This apparent contradiction has led researchers to propose the concept of the "double descent" risk curve, which extends and refines our understanding of the bias-variance tradeoff.

The double-descent curve describes a two-phase reduction in test risk as model complexity increases. The first descent (living in the under-parameterized regime with $d < d^*$) is the low complexity regime where models are underfitted, exhibiting high bias and low variance. As model complexity increases (d goes up), the bias decreases while variance increases, leading to a decrease in test error until the classical sweet spot d^* is reached. This sweet spot is located at the interpolation threshold ($d = d^*$), a critical point where the model has just enough complexity to fit the training data perfectly. At this point, the model is on the brink of overfitting.

The second descent (living in the over-parameterized regime with $d > d^*$) is the high complexity regime where models become over-parameterized, with more effective number of parameters than data points. Surprisingly, as model complexity continues to increase, test error decreases again. Now, the model begins to generalize better despite being able to fit the training data perfectly.

As illustrated in Figure 4.2, the relationship between model complexity and performance can exhibit an interesting pattern. When the effective model complexity is below the interpolation threshold, we observe the classical bias-variance tradeoff in action. This is characterized by an increasing test risk as

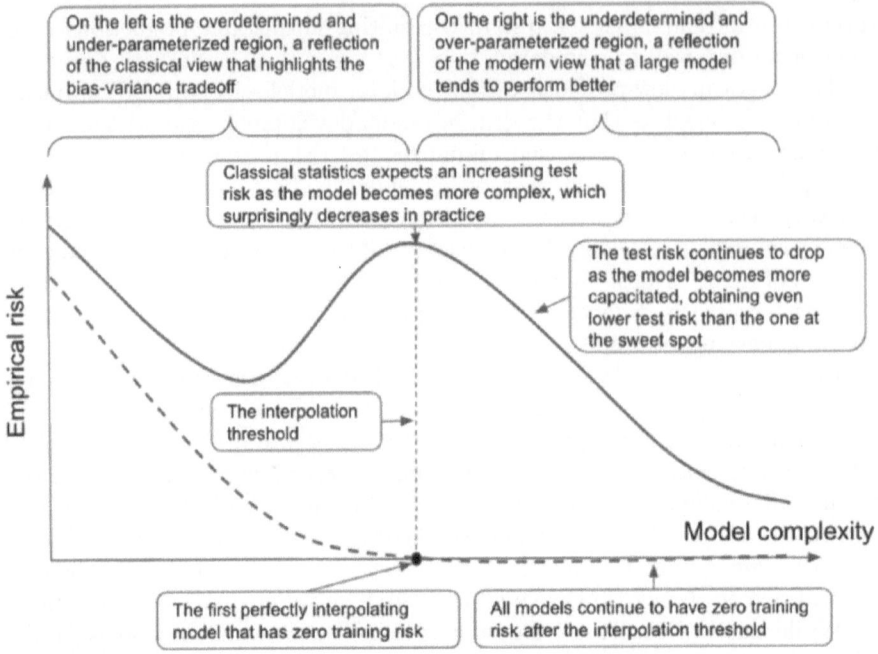

FIGURE 4.2
The "double-descent" curve that unifies the classical and modern views on generalization.

the model's representational capacity grows, reflecting the traditional understanding that overly complex models may overfit the training data. However, a remarkable shift occurs after passing the interpolation threshold: the test risk begins to decrease again. Throughout this entire process, the training risk consistently drops as the model complexity increases, eventually reaching and maintaining zero after surpassing the interpolation threshold. This behavior challenges our conventional understanding of model performance and generalization.

The "double-descent" curve in Figure 4.2 presents a unified perspective that bridges the gap between classical statistical learning theory and modern empirical observations in machine learning. This phenomenon has been observed across a diverse range of machine-learning models, from the polynomial regression discussed in the previous chapter to sophisticated deep learning architectures such as convolutional neural networks. In the under-parameterized region, as model complexity increases, we witness the first descent in test risk. This initial behavior aligns with the classical U-shaped curve that encapsulates the bias-variance tradeoff, where increasing complexity reduces bias but simultaneously raises variance. The model then reaches a critical point when its complexity becomes sufficient to interpolate the training data perfectly, marking the transition into the over-parameterized region.

Interestingly, it is in this over-parameterized regime that we observe behavior that defies traditional statistical intuition. As the model complexity continues to increase beyond the interpolation threshold, we see a second descent in test risk. This unexpected improvement in generalization performance challenges long-held beliefs about the relationship between model complexity and overfitting. It suggests that in certain scenarios, particularly with modern machine-learning models, increasing the model's capacity beyond the point of perfect training set fit can still lead to better generalization on unseen data. This phenomenon has sparked significant interest in the machine-learning community, prompting researchers to reevaluate our understanding of generalization and to develop new theoretical frameworks that can account for this behavior in over-parameterized models.

The "double-descent" phenomenon, despite its widespread occurrence, remains a largely unexplored and intriguing area of research in machine learning. The underlying mechanisms driving this observed behavior are not yet fully understood, presenting an exciting frontier for investigation. In the modern over-parameterized regime, we often encounter counterintuitive behaviors that challenge our traditional understanding of model performance. For example, deep neural networks frequently exhibit the "double-descent" phenomenon even in the absence of explicit regularization techniques, which turns out to be related to the way it is trained using the stochastic gradient descent (SGD) algorithm. Interestingly, when proper regularization methods such as L1 or L2 penalties are applied, the second descent may not necessarily manifest, suggesting that these regularization techniques can effectively mitigate overfitting even in highly complex models.

Furthermore, the modern interpolation regime presents another paradox: increasing the size of the training dataset can sometimes lead to a degradation in test performance. This observation stands in contrast to the conventional view that more data can invariably improve model learning. These interesting phenomena delineate the boundaries of classical statistical learning theory and underscore the need for deeper research into the statistical and computational principles governing modern machine-learning and deep learning models.

To further analyze the risk curve and gain a deeper understanding of its implications, we can consider each point on the curve as representing a fully trained model obtained through a specific training algorithm, such as SGD. When we examine a fixed level of model complexity, represented by a vertical slice in Figure 4.2, the training risk reflects the performance of the best-trained model on the training set for that particular complexity level. Similarly, the test risk denotes the performance of this best-trained model when applied to the test set. It is important to note that for any given level of complexity, multiple potential models can exist resulting from different initializations or optimization paths. The risk curve we observe thus summarizes the performance of the best-trained model at each level of complexity.

4.1.1 Beyond perfect interpolation

Complex models, such as neural networks and other nonlinear classes of models, can often exhibit remarkably low, or even zero, training error once surpassing the critical interpolating threshold. This phenomenon occurs when the model becomes sufficiently complex to interpolate the training data perfectly, as illustrated by the rightmost point in Figure 4.1. Such a situation can arise when the number of model parameters or weights, denoted as p, matches or exceeds the number of observations n in the training dataset. A classic example of this can be found in linear regression. When $p = n$, the problem essentially reduces to solve a system of n linear equations with n free variables. Under the condition of nonsingular matrix properties, this system has a unique solution that yields an exact fit to the training data. This scenario represents the point at which the model transitions from being under-parameterized to over-parameterized, marking a critical juncture in the model's capacity to fit the data.

To better illustrate this concept, let's consider a simple two-dimensional training set comprising two observation points, $\{x^{(1)}, y^{(1)}\}$ and $\{x^{(2)}, y^{(2)}\}$, on a coordinate system. If we start with a model that has only one parameter, such as a horizontal line represented by $f(x) = c$, we find that this univariate model is too simplistic to fit both points simultaneously. An exception occurs only if the two observations, $y^{(1)}$ and $y^{(2)}$, happen to have identical values. In this scenario, the system of equations is overdetermined and under-parameterized, with more constraints (equations) than free variables (parameters). Consequently, a solution may not exist if the points do not align horizontally, highlighting the limitations of an overly simplistic model.

The situation changes dramatically when we introduce two parameters, as in a simple linear regression model $f(x) = w_1 x + w_0$. This enhanced model renders the problem solvable with an exact solution. The two parameters, w_1 (slope) and w_0 (intercept), correspond to the two free variables in the system. We now have a system consisting of two equations, derived from the two observations $\{x^{(1)}, y^{(1)}\}$ and $\{x^{(2)}, y^{(2)}\}$. Solving this system determines a unique line that passes through both points of the coordinate system, resulting in zero training error. This case exemplifies an exactly determined system, where the number of constraints equals the number of parameters, allowing the model to interpolate the data perfectly and generate zero training error. It represents the precise point at which the model has just enough complexity to fit the training data without any excess capacity.

Now, extending the model further to include more than two parameters, such as in a polynomial model $f(x) = w_2 x^2 + w_1 x + w_0$, transforms the problem into an underdetermined and over-parameterized system. With three parameters and only two observations, the system no longer has a unique solution but instead admits infinitely many solutions. These solutions correspond to various curves that pass through both points, each characterized by different combinations of the weights w_2, w_1, and w_0. Such a model can interpolate

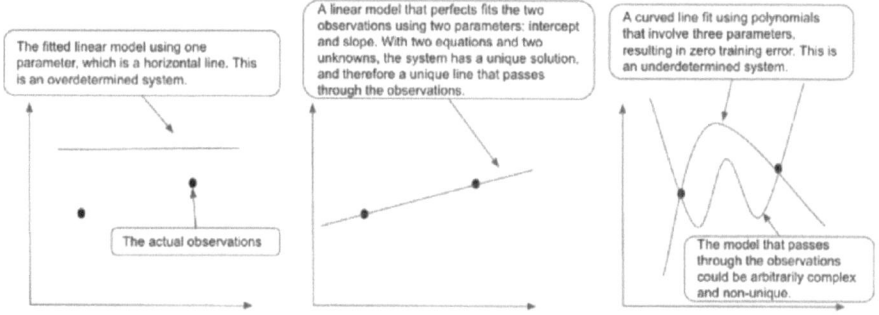

FIGURE 4.3
Fitting two two-dimensional observations using different model complexities.

the two points perfectly, maintaining zero training error, but the additional degrees of freedom allow the curve to become arbitrarily wiggly or complex, depending on the number of parameters involved. This scenario illustrates the potential for overfitting in over-parameterized models, where the model has the capacity to fit not only the underlying pattern but also any noise or idiosyncrasies in the training data. Understanding these dynamics is crucial for navigating the trade-offs between model complexity, fitting capacity, and generalization performance in many training scenarios.

This relationship between model complexity and training error can be summarized in Figure 4.3, which provides a visual representation of the transition between under-parameterized, exactly parameterized, and over-parameterized regimes. This figure offers insights into how the model's capacity to fit data evolves as its complexity increases.

In the left plot, we observe a 0th-degree polynomial model, which contains only one parameter. This simplistic model is represented by a horizontal line, illustrating the limitations of an under-parameterized system. With just one degree of freedom, the model struggles to capture the underlying pattern in the data, often resulting in a high training error. This scenario exemplifies the challenges faced when a model lacks sufficient complexity to learn the true relationship in the data.

The middle plot depicts a critical juncture where the number of model parameters exactly matches the number of observations. In this exactly parameterized regime, the solution becomes unique and exact. The model has just enough flexibility to fit all data points perfectly, resulting in zero training error. This scenario represents the ideal balance between model complexity and data fitting, where the model can capture the underlying pattern without introducing additional unnecessary complexity.

The right plot illustrates the over-parameterized regime, where the model contains more parameters than there are observations. In this scenario, the system becomes underdetermined, leading to non-unique solutions. The model now has the capacity to fit the data in infinitely many ways, often resulting in overly complex solutions. While these models can achieve zero training error, they may introduce unnecessary wiggles or complexity that doesn't reflect the true underlying relationship in the data (perturbed by potential noise).

Overall, as we move from left to right across Figure 4.3, we witness the model's evolution in its ability to fit the data. This progression highlights the transition from failing to fit the data properly (under-parameterized) to perfectly fitting the data (exactly parameterized) and finally to admitting an infinite number of overly complex solutions (over-parameterized). Understanding this relationship is crucial for model selection and interpretation, as it helps us navigate the trade-offs between model simplicity, data fitting, and generalization performance in various machine-learning applications.

4.1.2   Behind the double-descent phenomenon

Now let us go back to the conflict between theory and practice: despite the conventional view that excessive model complexity and near-perfect fit to training data lead to poor generalization, empirical observations have shown that increasing model complexity beyond the interpolation threshold can actually result in low error rates on test sets. This counterintuitive phenomenon has significantly influenced the training of deep neural networks in practice. A direct result is that practitioners often prefer sufficiently large neural networks to start with, as these models can easily interpolate the training dataset while still achieving good generalization on unseen test data if properly trained. What's even more remarkable is that this preference for large and complex architectures persists even when the training data contains high levels of noise, challenging our traditional understanding of overfitting and model selection.

The concept of the "double-descent" risk curve is a way to reconcile the observed phenomenon of over-parameterization with classical learning theory. Specifically, this new perspective extends the traditional U-shaped risk curve associated with the bias-variance tradeoff, incorporating what is now termed the modern interpolation regime. The double-descent framework introduces a critical point called the interpolation threshold, defined as the point at which the model achieves zero training error. For model capacities below this threshold, the test risk follows the classical U-shaped trajectory, reflecting the familiar balance between underfitting and overfitting. However, as the model capacity surpasses the interpolation threshold, we enter the over-parameterized regime where, contrary to classical expectations, the test risk begins to decrease again. This second descent in test risk characterizes the modern interpolation zone, challenging our traditional understanding of model complexity and generalization.

For now, let us use the number of parameters in the model as a proxy for its representational capacity, or complexity. A more complex model typically involves a greater number of parameters, allowing for more intricate representations of the data. As the model gets bigger and incorporates more parameters to the point that it surpasses the interpolation threshold, an interesting phenomenon occurs: the test risk declines even below the minimum risk observed at the "sweet spot" of the classical regime. This empirical result raises a great question that has captivated researchers and practitioners: why do larger, seemingly overfit models perform even better on unseen data, which stands in sharp contrast to our classical intuitions about generalization?

One good explanation for this phenomenon draws upon the principle of Occam's razor, a concept introduced in Chapter 1. This principle posits that among multiple models that fit the data equally well, we should prefer the simplest and smoothest model. In the context of machine learning, a common measure of smoothness is the norm of the model parameters. Models with smaller norms for their weights are generally considered more regularized and smoother, displaying an important inductive bias that encourages complex models to align with simple underlying structures. When multiple models achieve zero training error in the over-parameterized regime, the learning algorithm typically selects the one with the smallest norm in its weights. This selection process introduces a bias toward smooth solutions within the space of candidate functions that perfectly fit the training data, potentially explaining why over-parameterized models can still generalize well despite their high complexity.

In addition to parameter norms, various other techniques also play crucial roles in controlling the smoothness of a model. One notable example is SGD, which acts as an implicit regularizer by favoring smooth and low-variance models among those that perfectly interpolate the training data. This implicit regularization effect of SGD is particularly significant in the context of deep learning, where it contributes to the surprising generalization performance of neural networks despite their excessive capacity. Another important technique is bagging, or bootstrap aggregation, which can influence the smoothness of the fitted model by aggregating multiple model instances.

When faced with several models that yield zero training error, SGD demonstrates a tendency to prefer those with lower variance, which is often associated with smaller parameter norms. This preference becomes particularly important in the over-parameterized regime, where the function class is expansive. In this context, the bias toward smaller norms leads to the selection of "simpler" models that exhibit reduced variance in their predictions. The implicit regularization effect of SGD thus plays a crucial role in the generalization capabilities of deep neural networks, despite their seemingly excessive capacity. This phenomenon remains an active and exciting area of research, with ongoing investigations seeking to unravel the complex interplay between optimization algorithms and implicit regularization in deep learning.

Let us recall the definition of a norm, which provides a way to quantify the magnitude of a weight vector $\mathbf{w}$. One commonly used metric is the ℓ_p norm, defined for a weight vector $\mathbf{w}^\top = [w_1, w_2, \ldots, w_D]$ as

$$\ell_p = \left(\sum_{i=1}^{D} |w_i|^p \right)^{1/p}.$$

This general form gives rise to specific norms for different values of p. For instance, when $p = 1$, we obtain widely used ℓ_1 norm: $\ell_1 = |w_1| + |w_2| + \cdots + |w_D|$, which serves as the regularization penalty in Lasso regression. When $p = 2$, we get the ℓ_2 norm: $\ell_2 = \sqrt{w_1^2 + w_2^2 + \cdots + w_D^2}$, used as the regularization penalty in Ridge regression. Each of these norms imposes distinct regularization effects on a model, shaping its behavior and generalization properties in unique ways. The specific details and implications of these different norms will be explored more thoroughly in a later chapter dedicated to regularization via model cost.

It's crucial to understand that a regularized model doesn't necessarily need to achieve zero training error to generalize effectively to test data. In the realm of classical statistics, when a model exhibits signs of overfitting, the conventional recommendation is to employ regularization techniques. These techniques aim to reduce the model's capacity, guiding it toward the "sweet spot" on the classical U-shaped test risk curve. However, modern deep learning often adopts a seemingly counterintuitive approach: instead of reducing capacity, practitioners often increase model capacity by adding parameters. This strategy aims to escape the overfitting region by moving beyond the critical regime into the over-parameterized regime, which can offer an even lower test risk. This paradigm shift in approach highlights the evolving understanding of model complexity and generalization in the era of deep learning.

4.1.3 Extending the double-descent phenomenon via the scaling law

The double-descent risk curve has extended our notion of the bias-variance tradeoff. Historically, model complexity was primarily characterized by the size of the model, defined as the number of parameters used. However, subsequent research has revealed that the double-descent phenomenon also manifests as a function of training iterations in deep neural networks. Specifically, for a fixed model architecture, training with SGD demonstrates a double-descent pattern in test risk as the parameters gradually converge over training iterations. This observation has expanded our understanding of model behavior beyond simple parameter counts.

To unify these multiple perspectives on the double-descent phenomenon, we can use the concept of Effective Model Complexity (EMC) proposed by [14]. The EMC of a training procedure is defined as the maximum number of samples on which the procedure can achieve near-zero training error on average. For deep neural networks, a training procedure specifies both the model's size (or complexity) and the number of epochs.[1] This framework allows for a more nuanced understanding of model behavior, taking into account both structural and temporal aspects of training. Two types of double-descent curves have been empirically observed in the paper based on the training procedure: complexity-wise double descent, which arises when test risk exhibits a double-descent behavior as model complexity increases (i.e., as the number of parameters grows), and epoch-wise double descent, where for a fixed model complexity, test risk initially increases due to overfitting as training iterations proceed but subsequently decreases with further training epochs.

The relationship between EMC and the size of the training set provides a framework to categorize these double-descent curves into three distinct regimes, each with its own characteristics and implications for model performance. In the under-parameterized regime, when the EMC is significantly smaller than the size of the training set, the model underfits the data. Each feature captures only limited information about the underlying patterns, and increasing model complexity or training epochs reduces both the training and test risks. However, optimization in this regime is challenging due to the presence of numerous local minima, making convergence more difficult. This regime corresponds to models with insufficient capacity to represent the data effectively, often resulting in poor performance on both training and test sets.

When the EMC is significantly larger than the size of the training set, the model enters the over-parameterized regime. In this state, the model interpolates the training data, achieving near-zero training error. Interestingly, increasing model size or training epochs can continue to reduce test risk in this regime, contrary to classical expectations of overfitting. Optimization becomes relatively straightforward because there are many interpolating solutions, and some of these local minima could be equally good solutions compared with the optimal global minimum. This scenario is analogous to an underdetermined system of equations, where the number of unknowns exceeds the number of equations. The ease of optimization and the potential for continued improvement in test performance make this regime particularly interesting for deep learning practitioners, which essentially ushers in the new world of scaling law.

The critical regime occurs when the EMC is comparable to the size of the training set, placing the model in a critical state. This regime is characterized by its unpredictability: modifying the training procedure may either increase or decrease the test risk. The model's fit for individual features becomes less reliable, leading to more erratic behavior in generalization performance compared to the under- and over-parameterized regimes. This critical regime represents a transition point between underfitting and overfitting, where the model's behavior is most sensitive to small changes in parameters or training procedures. Understanding and navigating this regime is crucial for fine-tuning models and achieving better test performance.

---

[1] One epoch is a complete passé through the full training set.

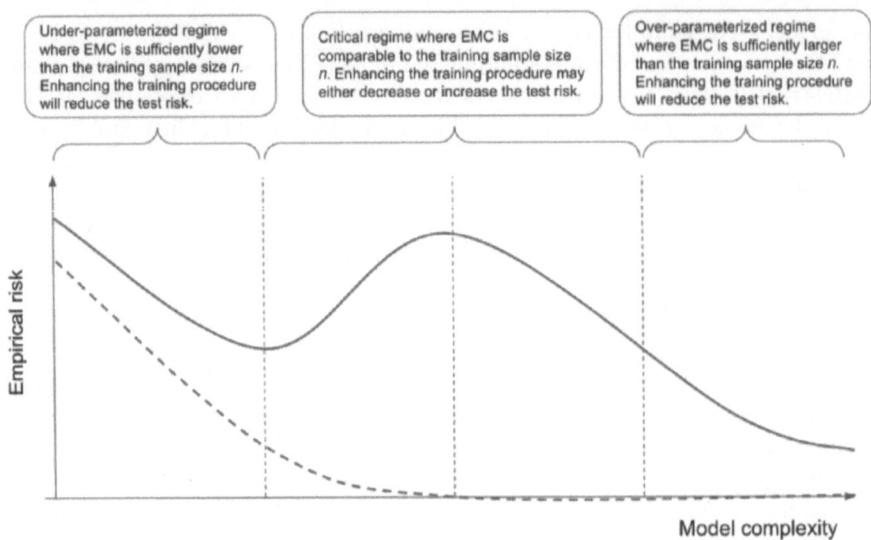

FIGURE 4.4
The zoning approach based on the generalized double-descent hypothesis.

This generalized double-descent hypothesis offers a formalization of the observations made in machine-learning and deep learning models regarding the relationship between model complexity and performance by zoning the learning curves into three distinct regimes: under-parameterized, critically parameterized, and over-parameterized. As illustrated in Figure 4.4, this new perspective provides a more nuanced understanding of how test risk evolves across different levels of model complexity and training procedures.

In this zoning approach, both the left and right zones represent areas where the test risk decreases as the training procedure is enhanced, such as by enlarging the model size or training for more epochs. These zones correspond to the under-parameterized and over-parameterized regimes, respectively. In the under-parameterized regime, models have low EMC relative to the size of the dataset, and increasing model complexity tends to reduce test risk. Similarly, in the over-parameterized regime, there is a monotonic relationship between EMC and test risk, with both improving as the training procedure is enhanced.

The middle zone, representing the critically parameterized regime, exhibits less predictable behavior. In this regime, changes to the training procedure can either increase or decrease the test risk, making it a challenging area for model optimization. This unpredictability stems from the model's EMC being comparable to the size of the training set, creating a delicate balance where small changes can have significant and sometimes counterintuitive effects on performance.

In the context of over-capacitated linear models, such as polynomial regression with excessive polynomial degrees, explicitly penalizing the parameter magnitudes introduces an inductive bias toward smoother functions, thereby improving generalization. This approach aligns with traditional regularization techniques used to combat overfitting. Similarly, in over-capacitated deep neural networks, the SGD algorithm plays a crucial role by implicitly regularizing the training process. SGD tends to favor low-variance models, which enhances generalization even in highly complex networks. This implicit regularization effect of SGD contributes to the surprising generalization capabilities of deep neural networks, despite their often vast parameter spaces.

To further explore and validate these concepts, we will examine the double-descent phenomenon in interpolating polynomial regression models with smoothness constraints in a later section. This analysis will provide empirical evidence demonstrating how the test risk evolves as the model becomes increasingly complex.

4.1.4 A brief history of the double-descent phenomenon

The double-descent phenomenon has historically been overlooked due to the typical workflows and assumptions surrounding classical model training. In earlier practices, linear and parametric models were designed with a fixed and limited capacity, constrained by a relatively small number of features. These models typically operated in settings where the number of features, p, was significantly smaller than the number of observations, n, in the dataset. As a result, classical models seldom reach the interpolation threshold where the double-descent phenomenon becomes observable. This limitation in model complexity, while practical for the computational resources of the time, inadvertently obscured the rich behavior that occurs beyond the interpolation threshold.

Non-parametric models, renowned for their flexibility and power, often employ smoothing techniques and regularization to control their complexity. Regularization, in particular, plays a crucial role in reducing the effective capacity of the model by penalizing the magnitude of model parameters. This approach effectively prevents the model from interpolating the training data, a key condition for observing double descent. The widespread use of regularization techniques in non-parametric models inadvertently suppressed the occurrence of the double-descent phenomenon, as these regularized models typically operate in a regime far from the interpolation threshold. This practice, while effective in preventing overfitting, also masked the potential benefits of over-parameterization that we now recognize.

The understanding of the double-descent phenomenon began to gain prominence through seminal contributions in recent years, marking a significant shift in our understanding of model behavior [3], formally introduced the concept of the interpolation regime. This significant work presented both theoretical and empirical evidence for the existence and ubiquity of the double-descent phenomenon across a broad spectrum of high-complexity models. The authors demonstrated that as model complexity increases beyond the interpolation threshold, test error often decreases again, contradicting the classical bias-variance intuition. This revelation challenged long-held beliefs about the relationship between model complexity and generalization performance, opening new avenues for research and model design.

Building upon this foundation, [14] further extended our understanding by studying double descent in deep neural networks. Their work provided compelling empirical evidence of the doubly descending risk curve as a function of both model size and the number of training iterations. These results showed that increasing model size, represented by the number of parameters p, can reduce test error even after reaching the interpolation threshold. Similarly, their research revealed that prolonged training could further reduce test error, highlighting the non-monotonic nature of risk reduction in over-parameterized regimes. These findings challenged the conventional wisdom that larger models and more training inevitably lead to overfitting, suggesting instead that there are benefits to pushing beyond traditional limits of model complexity.

In addition to these insights into model size and training duration, [14] further identified the intriguing phenomenon of sample non-monotonicity. They observed that, in certain cases, adding more training data to the dataset can paradoxically degrade test performance. This surprising result challenges the long-held belief that larger datasets invariably improve generalization. The discovery of sample non-monotonicity underscores the complex interplay between model capacity, dataset size, and generalization performance, suggesting that the relationship between these factors is more nuanced than previously thought. This finding has significant implications for data collection and model training strategies in machine learning.

The emergence of these works has fundamentally reshaped our understanding of the bias-variance trade-off in modern machine learning. By highlighting the critical role of the interpolation threshold and over-parameterized regimes, they have bridged the gap between classical statistical paradigms and contemporary deep learning practices, providing a unified perspective on the dynamics of model generalization. This new understanding has not only reconciled seemingly contradictory observations in machine-learning practice but has also opened up new research directions and practical strategies for model design and optimization. As we continue to explore the implications of double descent and related phenomena, we are likely to uncover even more insights that will further refine our approach to building and training machine-learning models.

4.2   Double Descent in Polynomial Regression

In the previous chapter, we explored the task of modeling a nonlinear relationship between a univariate predictor in X and an outcome variable y using polynomial regression. This approach allows us to capture complex, non-linear patterns in data. By expressing y as a function of X with an additive error term, the model's prediction takes the form $\hat{y} = f_w(X)$, where $f_w(X)$ represents the polynomial regression function parameterized by weights w. This formulation provides a flexible framework for modeling a wide range of relationships between variables.

Specifically, the polynomial regression function $f_w(X)$ is defined as $f_w(X) = \Phi(X)w$, where $\Phi(X)$ denotes the design matrix for polynomial features of X, and w is the corresponding vector of coefficients. Expanding this expression, we can write the function as a sum of polynomial terms:

$$f_w(X) = \sum_{i=0}^{p} w_i X^i = w_0 + w_1 X + w_2 X^2 + \cdots + w_p X^p.$$

This formulation allows for great flexibility in modeling, as the degree of the polynomial can be adjusted to capture increasingly complex relationships. For example, if we restrict the degree of the polynomial to $p = 2$, the regression model becomes:

$$f_w(X) = w_0 + w_1 X + w_2 X^2,$$

which captures quadratic relationships between the predictor X and the response variable y. This quadratic model can represent a variety of nonlinear patterns, including parabolic relationships that are common in many real-world phenomena.

Training a polynomial regression model involves estimating the weight vector w that minimizes a chosen loss function, typically the mean squared error, over the training dataset $\{(x^{(i)}, y^{(i)})\}_{i=1}^{n}$. This process aims to find the optimal coefficients that best fit the observed data. However, as we increase the degree of the polynomial, p, we encounter both opportunities and challenges. A higher polynomial degree results in a more flexible model capable of capturing intricate patterns in the data, potentially reducing bias and improving the fit to the training data. However, this increased flexibility comes at a cost: it also leads to a higher risk of overfitting, characterized by a wiggly fit and high variance. This high variance arises because the model becomes overly sensitive to small fluctuations in the training data, potentially capturing noise rather than underlying patterns.

The challenge, therefore, lies in striking a delicate balance: we want to accurately represent the training data while keeping the model smooth and simple enough to avoid overfitting and ensure good generalization to unseen data. This balance is crucial for creating models that are not only accurate on the training set but also perform well on new, unseen data, a fundamental goal in machine-learning and statistical modeling.

To achieve this balance, smoothing splines offer an effective and elegant solution. A smoothing spline introduces an explicit smoothness constraint during the fitting process. This constraint penalizes the complexity of the model by discouraging large fluctuations in the fitted curve, effectively controlling its flexibility. Mathematically, this is achieved by minimizing a loss function that incorporates both the goodness-of-fit to the training data and a penalty term that enforces smoothness. This approach allows us to use high-degree polynomials while mitigating the risk of overfitting, providing a powerful tool for nonlinear regression.

The smoothing spline framework can reveal the double-descent phenomenon. By controlling the smoothness of the polynomial regression model, smoothing splines allow for high-degree polynomials to generalize well without succumbing to the adverse effects of high variance. This ability to maintain good performance even with highly complex models is key to understanding and leveraging the double-descent effect. In the next section, we will delve deeper into the mathematical formulation of smoothing splines and demonstrate their role in mitigating the double-descent effect. This exploration will provide valuable insights into how we can design and train models that achieve excellent performance across a wide range of complexities.

4.2.1   The smoothing spline

The term "spline" has its roots in a tool historically used by draftsmen and shipbuilders to create smooth curves. In classical statistics, splines have been widely employed to construct flexible functional forms, providing a powerful method for modeling complex relationships in data. The process of fitting a spline typically begins by identifying knots, which are points where two smooth curves join. The challenge then lies in fitting a curve between consecutive knots that not only captures the underlying pattern in the data but also maintains smoothness and continuity across the entire range.

To ensure the smoothness and continuity of the resulting curve, statisticians impose constraints on the derivatives of a specific order. A common and particularly effective approach is to utilize the second derivative of the fitted function and penalize the integrated squared second derivatives. This penalty serves a dual purpose: it ensures that the curve is not overly rough or wiggly, while simultaneously minimizing the residual sum of squares (RSS) between the model predictions and the target outcomes. This balance between smoothness and fit is at the heart of the spline approach.

Formally, we can define the penalty on the second derivative of the function f_w as:

$$J(w) = \int \left(f_w''(x) \right)^2 dx.$$

This penalty term $J(w)$ plays a crucial role in regulating the curvature of the model, effectively discouraging excessive fluctuations that might represent noise rather than true underlying patterns. In the context of smoothing splines, the resulting model seeks to minimize a cost function that incorporates

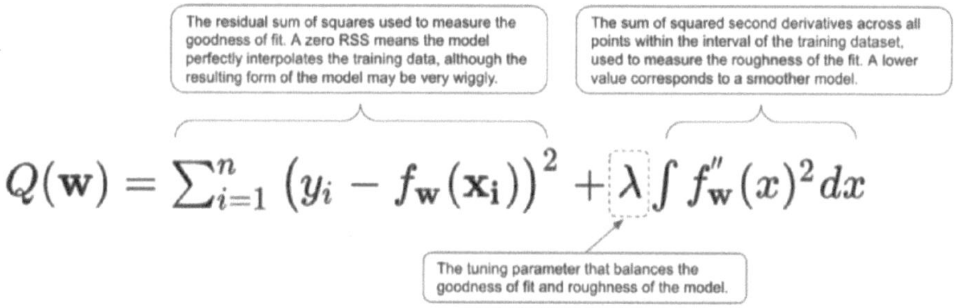

The residual sum of squares used to measure the goodness of fit. A zero RSS means the model perfectly interpolates the training data, although the resulting form of the model may be very wiggly.

The sum of squared second derivatives across all points within the interval of the training dataset, used to measure the roughness of the fit. A lower value corresponds to a smoother model.

$$Q(\mathbf{w}) = \sum_{i=1}^{n} \left(y_i - f_{\mathbf{w}}(\mathbf{x_i})\right)^2 + \lambda \int f_{\mathbf{w}}''(x)^2 dx$$

The tuning parameter that balances the goodness of fit and roughness of the model.

FIGURE 4.5
Penalized cost function for the smoothing spline model.

both the fit to the data and this smoothness penalty:

$$Q(w) = \sum_{i=1}^{n} \left(y^{(i)} - f_w(x^{(i)})\right)^2 + \lambda \int \left(f_w''(x)\right)^2 dx,$$

where λ is a weighting factor, or regularization parameter, that determines the tradeoff between the fidelity to the training data and the roughness of the curve. The selection of this parameter λ is typically performed through cross-validation, as its value critically influences the behavior of the resulting model.

The role of λ in shaping the model's behavior is profound and nuanced. A larger λ enforces greater smoothness by heavily penalizing high curvature, resulting in a smoother functional form that may sacrifice some fit to the data in favor of simplicity. Conversely, a smaller λ allows for more curvature, potentially leading to a rougher curve that better fits the individual data points but may be more prone to overfitting. When $\lambda \to 0$, the penalty term effectively vanishes, and the cost function reduces to the RSS. In this scenario, the model essentially becomes an interpolating spline, which perfectly fits the training data but may take on a highly wiggly shape that poorly generalizes to new data. At the other extreme, as $\lambda \to \infty$, the roughness penalty dominates the cost function, constraining the model to be a linear function, as this minimizes the second derivative everywhere. Between these extremes, λ serves as a critical hyperparameter that governs the balance between overfitting and underfitting, allowing the model to capture complex patterns while maintaining reasonable smoothness.

Figure 4.5 provides a visual illustration of these opposing forces in the penalized cost function. The RSS component promotes fitting the training data closely, while the roughness penalty encourages smoothness in the functional form.

The second derivative, $f_w''(x)$, plays a central role in this framework by measuring the rate of change of the slope of the fitted curve. It provides a quantitative measure of the roughness of the curve, with higher values of $f_w''(x)$

corresponding to more abrupt changes in curvature. By incorporating $J(w)$ into the cost function, the smoothing spline framework can avoid the need to predefine the number of knots in the model, a task that can be challenging and somewhat arbitrary. Instead, the degree of smoothness is automatically determined by the balance dictated by λ. A high penalty discourages roughness and promotes smooth curves, while a low penalty allows the model to capture more complex patterns in the data, adapting to the underlying structure without requiring manual specification of the model's flexibility.

This approach exemplifies how statistical learning can achieve a delicate balance between model complexity and generalization by introducing principled constraints on the functional form. The smoothing spline framework demonstrates that by carefully controlling the smoothness of the fitted function, we can create models that are both flexible enough to capture complex relationships and constrained enough to avoid overfitting.

4.2.2 Rewriting the smoothing spline cost function

Rewriting the smoothing spline cost function establishes a critical connection between classical linear regression, ridge regression, and smoothing splines, illuminating the evolution of regularization techniques in statistical modeling. Recall that the closed-form solution for linear regression is derived by setting the gradient of the cost function $Q(\mathbf{w})$ to zero and solving a system of normal equations:

$$\Phi^T \Phi \mathbf{w} = \Phi^T y.$$

Here, Φ denotes the design matrix (which could be a transformed and augmented version of the original design matrix X), $\mathbf{w}$ is the vector of weights, and y represents the target outcomes. This fundamental equation forms the basis for more advanced regularization techniques. When a new penalty term $J(\mathbf{w})$ is introduced, the derivation proceeds by incorporating the first-order gradient, leading to a more generalized normal equation that balances data fit with model complexity.

Using the same matrix notation, we can express the prediction function for the polynomial regression model as $f_{\mathbf{w}} = \Phi\mathbf{w}$. This formulation allows us to reformulate the cost function in a more compact and insightful manner:

$$Q(\mathbf{w}) = \|y - \Phi\mathbf{w}\|^2 + \lambda\mathbf{w}^T \Omega\mathbf{w}.$$

In this expression, Φ represents the list of polynomial basis functions used to construct the design matrix, and each entry $(\Phi)_{jk}$ is defined as $\phi_k(x^{(j)}) = (x^{(j)})^k$, where ϕ_k denotes the kth polynomial basis function applied to the jth input $x^{(j)}$. The double vertical bars $\| \cdot \|^2$ indicate the squared norm, which quantifies the sum of squared residuals between the predictions $\Phi\mathbf{w}$ and the target vector y. The matrix Ω plays a crucial role in this formulation, encoding the integrated second derivatives of the basis functions and acting as the roughness penalty. Formally, the entries of Ω are given by:

$$(\Omega)_{jk} = \int \phi_j''(x)\phi_k''(x)\,dx,$$

where $\phi_j''(x)$ and $\phi_k''(x)$ denote the second derivatives of the jth and kth basis functions, respectively. This integral form of Ω captures the essence of the smoothness constraint, penalizing rapid changes in the curvature of the fitted function.

This formulation bears a striking resemblance to the cost function of ridge regression but with a crucial distinction: the introduction of the Ω term in the penalty. In ridge regression, the penalty term is proportional to the squared norm of the weights, leading to normal equations of the form:

$$(\Phi^T\Phi + \lambda I)\mathbf{w} = \Phi^T y,$$

where I is the identity matrix, a diagonal matrix with ones along the main diagonal and zeros elsewhere. This form of regularization uniformly penalizes large weight values. For smoothing splines, however, the generalized normal equations take on a more sophisticated form:

$$(\Phi^T\Phi + \lambda\Omega)\mathbf{w} = \Phi^T y.$$

The additional Ω term accounts for the second-derivative roughness penalty, making the smoothing spline formulation more general and flexible than ridge regression. This generalization allows for a more nuanced control over the smoothness of the fitted function. In fact, when $\Omega = I$, the smoothing spline cost function reduces to the ridge regression cost function, highlighting the relationship between these regularization techniques.

Figure 4.6 provides a comparison of linear regression, ridge regression, and smoothing splines by highlighting the role of the penalty term, the formulation of the cost function, and the resulting normal equations under different penalization schemes. This comparison underscores the evolutionary nature of these regularization techniques, from the simplicity of linear regression to the sophisticated smoothness control of smoothing splines. Notably, the smoothing spline model extends ridge regression by incorporating an additional term that penalizes the roughness of the fitted function. This penalty ensures that the resulting function is not only well-fitted to the data but also smooth and continuous, striking a balance between fidelity to the data and the underlying assumption of smoothness in the true relationship.

4.2.3 Deriving the closed-form solution

To derive the closed-form solution for the smoothing spline problem, we extend the concept of normal equations used in linear regression by introducing an augmented design matrix and an additional vector. This approach provides a general framework for incorporating the smoothness penalty into the

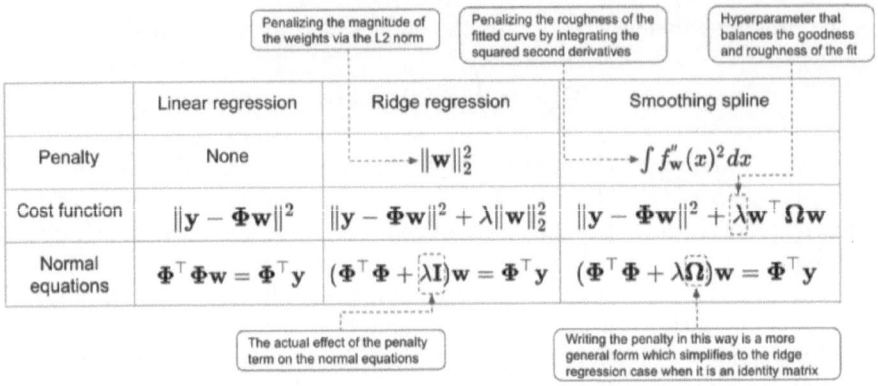

FIGURE 4.6
Summarizing the penalty term, cost function, and normal equations under different penalization schemes.

regression model. The augmented design matrix, denoted as Φ_*, combines the original design matrix Φ with additional rows contributed by the penalty term. Formally, this is expressed as:

$$\Phi_* = \begin{pmatrix} \Phi \\ \sqrt{\lambda}B \end{pmatrix},$$

where B is defined such that $B^T B = \Omega$, and Ω is the previous $p \times p$ symmetric penalty matrix encoding the second-derivative roughness penalties. By scaling Ω by $\sqrt{\lambda}$, we ensure that the penalty term $\lambda \mathbf{w}^T \Omega \mathbf{w}$ is properly incorporated into the augmented matrix Φ_*. This augmentation effectively concatenates p additional rows to Φ, with $\sqrt{\lambda}B$ contributing the rows associated with the penalty term. This construction allows us to integrate the smoothness constraint into the regression framework.

Similarly, the output vector y is extended by appending p zeros to the end, forming the augmented output vector:

$$y_* = \begin{pmatrix} y \\ 0 \end{pmatrix}.$$

This augmentation of the output vector complements the extended design matrix, ensuring that the dimensions of the system remain consistent while incorporating the penalty term.

With the augmented design matrix Φ_* and output vector y_*, we can construct the normal equations for the augmented system. Based on the original normal equations $\Phi \mathbf{w} = y$, the augmented equations are:

$$\Phi_* \mathbf{w} = y_*$$

Multiplying both sides by Φ_*^T, we obtain:

$$\Phi_*^T \Phi_* \mathbf{w} = \Phi_*^T y_*.$$

This step is crucial as it transforms the problem into a form that can be solved using standard linear algebra techniques. Substituting the expressions for Φ_* and y_*, this expands to:

$$\left(\Phi^T \quad \sqrt{\lambda}B^T\right)\begin{pmatrix}\Phi \\ \sqrt{\lambda}B\end{pmatrix}\mathbf{w} = \left(\Phi^T \quad \sqrt{\lambda}B^T\right)\begin{pmatrix}y \\ 0\end{pmatrix}.$$

Carrying out the matrix multiplications yields:

$$(\Phi^T \Phi + \lambda B^T B)\mathbf{w} = \Phi^T y.$$

Since $B^T B = \Omega$, this simplifies to:

$$(\Phi^T \Phi + \lambda\Omega)\mathbf{w} = \Phi^T y.$$

This result introduces the penalty term $\lambda\Omega$ into the normal equations, accounting for the roughness constraints imposed by the second-derivative penalty. The resulting equation provides a balanced solution that considers both the fit to the data and the smoothness of the function.

It turns out that the cost function can also be reformulated using the augmented variables Φ_* and y_*, as follows:

$$Q(\mathbf{w}) = \|y - \Phi\mathbf{w}\|^2 + \lambda\mathbf{w}^T \Omega\mathbf{w} = \|y_* - \Phi_*\mathbf{w}\|^2.$$

This reformulation demonstrates how the augmented system encapsulates both the data-fitting term and the smoothness penalty in a single squared norm expression. Setting the gradient of $Q(\mathbf{w})$ with respect to $\mathbf{w}$ to zero leads to the same normal equations:

$$\Phi_*^T \Phi_* \mathbf{w} = \Phi_*^T y_*.$$

This derivation highlights the role of the augmented design matrix and output vector in incorporating the penalty term into the least squares framework. The augmented design matrix Φ_* enlarges the column space of Φ from $\mathbb{R}^n$ to $\mathbb{R}^{n+p}$, which can stabilize the solution of the normal equations. This expansion of the column space is particularly beneficial when dealing with ill-conditioned or near-singular design matrices. Additionally, the matrix Ω often undergoes transformations, such as eigendecomposition, to ensure that its column vectors are mutually orthogonal, further mitigating issues of collinearity.

An alternative derivation of the normal equations uses the first-order optimality condition, providing a different perspective on the problem. Taking the gradient of $Q(\mathbf{w})$ with respect to $\mathbf{w}$ and setting it to zero gives:

$$\frac{\partial Q(\mathbf{w})}{\partial \mathbf{w}} = \frac{\partial}{\partial \mathbf{w}}\left((y-\Phi\mathbf{w})^T(y-\Phi\mathbf{w})+\lambda\mathbf{w}^T\Omega\mathbf{w}\right) = -2\Phi^T y+2\Phi^T \Phi\mathbf{w}+2\lambda\Omega\mathbf{w} = 0.$$

Rearranging terms yields:

$$(\Phi^T \Phi + \lambda \Omega)\mathbf{w} = \Phi^T y.$$

While both approaches lead to the same normal equations, the augmented matrix approach provides deeper insights into the geometric interpretation of the penalty term. By expanding the column space of Φ, the augmented system ensures greater stability of the solution, especially for large λ. Furthermore, as λ increases, the extended column vectors in Φ_* become more orthogonal, enhancing numerical stability and robustness in solving the normal equations. This geometric perspective underscores the value of the augmented formulation in addressing challenges such as collinearity and overfitting in the original design matrix Φ. The augmented approach not only provides a mathematically elegant solution but also offers practical benefits in terms of computational stability and interpretability of the smoothing spline model.

4.2.4   Implementing the smoothing spline model

To demonstrate the double-descent phenomenon using the smoothing spline model, we begin by defining a true underlying function to approximate. This function, which will be unknown to the machine learner, represents a piecewise linear form given by:

$$y = \|\|x - 0.4\| - 0.2\| + \frac{x}{2} - 0.1,$$

where x represents the single input variable. The function includes absolute value transformations, ensuring piecewise linear behavior across different regions of the input space. See the following code listing for its definition.

```
1  # Function to represent the true underlying piecewise linear
        function
2  def get_true_y(x):
3      return np.abs(np.abs(x - 0.4) - 0.2) + x / 2 - 0.1
```

Listing 4.1
True Underlying Function.

Next, we generate equally spaced input points across the range $[0, 1]$ to evaluate and plot out the corresponding function values via a line chart.

```
1   # Import necessary libraries
2   import math
3   import matplotlib.pyplot as plt
4   import numpy as np
5   from numpy import random
6
7   # Set seed to ensure reproducibility
8   random.seed(1)
9
10  # Generate evenly spaced input values
```

```
11  X_vals = np.linspace (0, 1, 100).reshape (-1, 1)
12
13  # Compute the true function values
14  y_vals = get_true_y (X_vals)
```

Listing 4.2
Generating Functional Evaluations.

Our goal is to approximate this function using the smoothing spline model with a varying number of polynomial basis terms. We first implement a transformation function to map the original univariate inputs x into a new design matrix Φ, where each column corresponds to a higher-degree polynomial term:

$$\Phi_{jk} = \phi_k(x^{(j)}) = (x^{(j)})^k, \quad k \in \{0, 1, \ldots, d\},$$

where d denotes the maximum degree of the polynomial.

```
1   # Function to generate the new design matrix with polynomial
        basis
2   def polynomial (x, num_basis=2):
3       # Create a zero-initialized design matrix
4       Phi = np.zeros ((x.shape[0], num_basis))
5
6       # Fill in each column with the corresponding polynomial term
7       for i in range (num_basis):
8           Phi[:, i:i+1] = x ** i
9
10      return Phi
```

Listing 4.3
Generating Polynomial Function.

Now for the roughness penalty $J(\mathbf{w})$, it can approximated using the matrix Ω such that:

$$J(\mathbf{w}) = \mathbf{w}^T \Omega \mathbf{w}, \quad \text{where } \Omega_{qr} = \frac{(q-1)q(r-1)r}{q+r-3}.$$

To derive the penalty matrix Ω, we start from the roughness penalty term in the smoothing spline cost function:

$$J(\mathbf{w}) = \int \left(f_{\mathbf{w}}''(x) \right)^2 dx.$$

In the matrix-based implementation, this integral is approximated using the quadratic form:

$$J(\mathbf{w}) = \mathbf{w}^T \Omega \mathbf{w},$$

where Ω is the penalty matrix that encodes the contributions of the polynomial basis functions to the roughness penalty. The smoothing spline uses a polynomial basis, and the second derivative of the k-th basis function $\phi_k(x)$ is given by:

$$\phi_k''(x) = k(k-1)x^{k-2}, \quad k \geq 2.$$

The integral for $J(\mathbf{w})$ can then be expanded as a double summation over all pairs of polynomial basis functions:

$$J(\mathbf{w}) = \sum_{q=2}^{p} \sum_{r=2}^{p} \beta_{qr} w_q w_r,$$

where β_{qr} represents the interaction between the q-th and r-th basis functions. Each β_{qr} is computed as:

$$\beta_{qr} = \int \phi_q''(x) \phi_r''(x) \, dx.$$

Using the formula for the second derivative of the polynomial basis function, we have:

$$\phi_q''(x) = q(q-1)x^{q-2}, \quad \phi_r''(x) = r(r-1)x^{r-2}.$$

The product of the second derivatives is:

$$\phi_q''(x)\phi_r''(x) = q(q-1)r(r-1)x^{q+r-4}.$$

The integral over the interval $[0, 1]$ is then:

$$\beta_{qr} = \int_0^1 q(q-1)r(r-1)x^{q+r-4} \, dx.$$

The integral of x^{q+r-4} is calculated as:

$$\int_0^1 x^{q+r-4} \, dx = \frac{1}{q+r-3}, \quad q+r > 3.$$

Substituting this result, we get:

$$\beta_{qr} = \frac{q(q-1)r(r-1)}{q+r-3}.$$

Now we can use this closed-form solution to calculate the coefficients of the model.

```
# Function to estimate coefficients of the smoothing spline model
def estimate_coef(X, y, num_basis):
    # Generate the polynomial design matrix
    Phi = polynomial(X, num_basis=num_basis)

    # Define the penalty matrix omega
    q = np.arange(2, num_basis).reshape(1, -1)
    r = q.reshape(-1, 1)
    omega = np.zeros((num_basis, num_basis))
    omega[2:, 2:] = (q - 1) * q * (r - 1) * r / (q + r - 3)

    # Perform eigendecomposition for stability
    values, vectors = np.linalg.eig(omega)
    beta = vectors @ np.diag(values.clip(min=0) ** 0.5)
```

```
16    # Construct augmented matrices for the normal equations
17    y_star = np.concatenate((y, np.zeros(omega.shape[0]).reshape
          (-1, 1)), axis=0)
18    Phi_star = np.concatenate((Phi, math.sqrt(rho) * beta.T),
          axis=0)
19
20    # Solve the normal equations using least squares
21    w_star = np.linalg.lstsq(Phi_star, y_star, rcond=None)[0]
22
23    return w_star
```

Listing 4.4
Obtaining Closed-form Solution.

To evaluate the model performance, we calculate the empirical risk using the MSE:

$$Q = \frac{1}{n}\|y - \Phi\mathbf{w}\|^2,$$

where Φ is the design matrix and $\mathbf{w}$ represents the estimated coefficients.

```
1    # Function to compute mean squared error
2    def calculate_mse(X, y, coef, num_basis):
3        # Expand the original design matrix
4        Phi = polynomial(X, num_basis=num_basis)
5
6        # Compute the model predictions
7        f_star = Phi @ coef
8
9        # Compute residuals and the MSE
10       resid = y - f_star
11       Q = np.dot(resid.T, resid).mean()
12
13       return Q
```

Listing 4.5
Calculating MSE.

Finally, we run multiple simulations to compute the median training and test MSE for different polynomial degrees by running different simulations. The double-descent phenomenon is examined by plotting the training and test risks against model complexity.

```
1    # Function to compute performance over multiple runs
2    def compute_performance(num_train_samples, num_runs,
         num_basis_max, train_noise_std=0):
3        # Placeholder matrices for storing MSE
4        mse_train = np.zeros((num_runs, num_basis_max))
5        mse_test = np.zeros((num_runs, num_basis_max))
6
7        for k in range(num_runs):
8            # Generate training data
9            X_train = random.rand(num_train_samples).reshape(-1, 1)
10           y_train = get_true_y(X_train).reshape(-1, 1)
11
```

```
12          # Add noise to the training targets if specified
13          if train_noise_std > 0:
14              y_train += np.random.normal(0, train_noise_std,
                    y_train.shape[0]).reshape(-1, 1)
15
16          # Generate test data
17          X_test = np.linspace(0, 1, 100).reshape(-1, 1)
18          y_test = get_true_y(X_test).reshape(-1, 1)
19
20          # Assess models with varying polynomial degrees
21          for num_basis in range(1, num_basis_max + 1):
22              fitted_coef = estimate_coef(X_train, y_train,
                    num_basis)
23              mse_train[k, num_basis-1] = calculate_mse(X_train,
                    y_train, fitted_coef, num_basis)
24              mse_test[k, num_basis-1] = calculate_mse(X_test,
                    y_test, fitted_coef, num_basis)
25
26      return np.median(mse_train, axis=0), np.median(mse_test, axis
            =0)
27
28  # Global parameters
29  lamb = 1e-12
30  num_runs = 10
31  num_train_samples = 10
32  num_basis_max = 30
33
34  # Plot training and test MSE
35  fig = plt.figure()
36  ax = fig.add_subplot(1, 1, 1)
37  ax.set_yscale('log')
38  ax.set_xlabel('Polynomial Degree', labelpad=10)
39  ax.set_ylabel('MSE', labelpad=10)
40  ax.axvline(x=num_train_samples, color='gray', linewidth=0.5,
            linestyle='--')
41  ax.text(num_train_samples - 0.2, 1e-16, 'Interpolation Threshold'
            , fontsize=10, color='gray', rotation=90)
42
43  mse_train, mse_test = compute_performance(num_train_samples,
            num_runs, num_basis_max)
44  ax.plot(np.arange(1, num_basis_max + 1), mse_train, color='blue',
            linestyle='--', label='Train')
45  ax.plot(np.arange(1, num_basis_max + 1), mse_test, color='red',
            label='Test')
46  ax.legend(frameon=False)
47  plt.show()
```

Listing 4.6
Median MSE for Different Degrees of Polynomials.

Figure 4.7 demonstrates the empirical double-descent phenomenon. As expected, the training curve displays a monotonically decreasing trend as the model becomes more complex. The test risk initially decreases with increasing model complexity, peaks near the interpolation threshold, and then decreases again as complexity continues to grow. This behavior highlights the unique

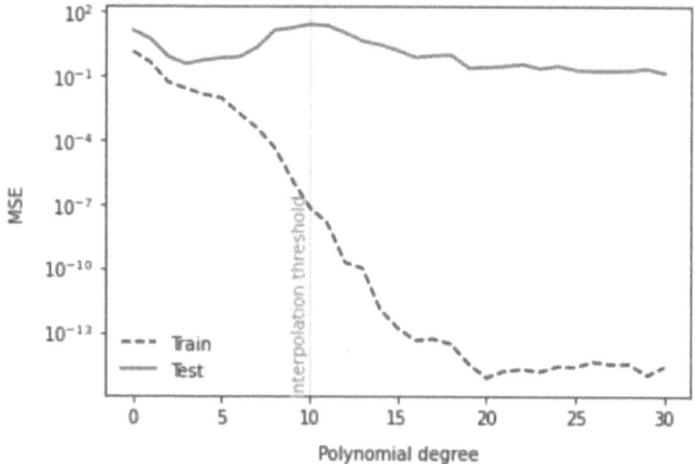

FIGURE 4.7
Median MSE against different degrees of polynomials.

generalization capabilities of over-parameterized models in modern machine learning. In this case, we managed to reproduce such a phenomenon with the help of a smoothness penalty in polynomial regression.

4.2.5 Observing goodness and roughness of fit with polynomial degrees

To examine the relationship in model complexity (measured by the degree of the polynomial) and the tradeoff between goodness and roughness of fit, we fit the smoothing spline model with varying polynomial degrees. The following code generates the training dataset, computes the true function values, and visualizes the fitted model for various polynomial degrees and the resulting goodness of fit. Each plot shows the true function, training points, and the fitted polynomial curve.

```
1  # Generate the training set
2  X_train = random.rand(num_train_samples).reshape(-1, 1)
3  y_train = get_true_y(X_train).reshape(-1, 1)
4
5  # Generate test data points along the true function
6  X_test = np.linspace(0, 1, 100, dtype=X_train.dtype).reshape(-1,
       1)
7  y_test = get_true_y(X_test).reshape(-1, 1)
8
9  # Fit and visualize the model for different polynomial degrees
10 for num_basis in range(1, num_basis_max + 1):
11     fig = plt.figure()
12     ax = fig.add_subplot(1, 1, 1)
13     ax.set_title(f'Degree {num_basis - 1}')
```

```
14    ax.plot(X_test, y_test, color='black', label='True function',
          linestyle='--')  # Plot true function
15    ax.scatter(X_train, y_train, color='blue', label='Train
          samples')  # Plot training samples
16
17    # Estimate model coefficients
18    fitted_coef = estimate_coef(X_train, y_train, num_basis)
19
20    # Generate predictions from the model
21    Phi = polynomial(X_test, num_basis=num_basis)
22    f_star = Phi @ fitted_coef
23
24    # Plot the fitted polynomial
25    ax.plot(X_test, f_star, color='red', label='Fitted polynomial
          ')
26    ax.legend(frameon=False)
27    plt.show()
```

Listing 4.7
Assessing Goodness of Fit with Varying Degrees of Polynomials.

Figure 4.8 shows the goodness of fit at different degrees of polynomials. When the model uses up to the 1st degree of polynomial (two features, including the constant term), the fitted model is a straight line:

$$f_{\mathbf{w}}(x) = w_0 + w_1 x.$$

As we can see, the straight line does not capture the piecewise linear behavior of the true function. This shows underfitting, as the model lacks the

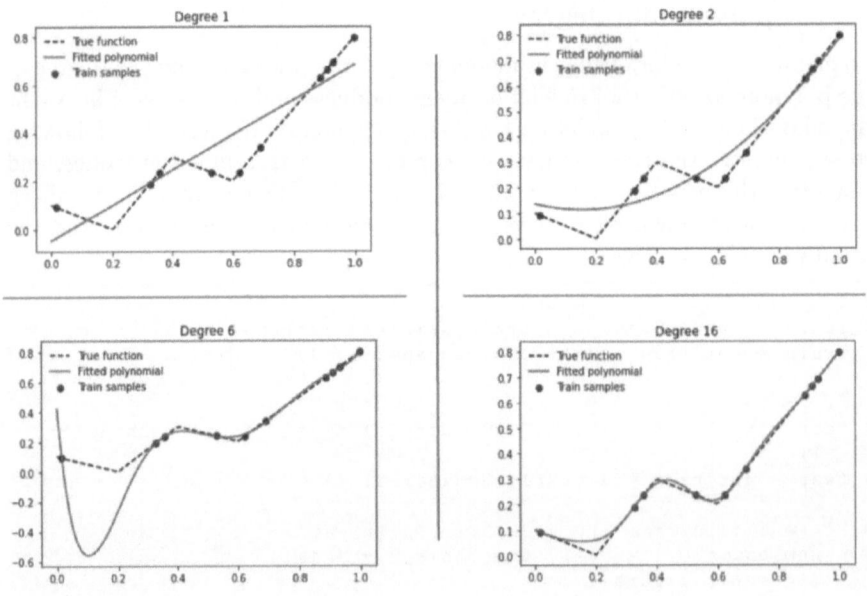

FIGURE 4.8
Visualizing the goodness of fit at different degrees of polynomials.

flexibility needed to approximate the true relationship. Increasing the degree of the polynomial is necessary to reduce the bias of the model.

With up to the 2nd degree of polynomial (three features), the fitted model gains the ability to represent nonlinear relationships:

$$f_\mathbf{w}(x) = w_0 + w_1 x + w_2 x^2.$$

As shown in Figure 4.8, the quadratic term introduces curvature to the fitted model, allowing it to approximate nonlinear patterns in the data. However, this curvature is still insufficient to closely follow the true function's piecewise linear structure, indicating the need for higher-order polynomials to improve the fit.

Now, when the model includes up to the 6th degree of polynomial, it becomes flexible enough to closely approximate the piecewise linear structure of the true function:

$$f_\mathbf{w}(x) = w_0 + w_1 x + w_2 x^2 + \cdots + w_6 x^6.$$

The fitted model accurately interpolates the training points. However, the model displays high variance in regions where training points are sparse, as evidenced by the steep dip between the first and second points. This is a classical overfitting scenario, where the model becomes overly sensitive to noise in the training data.

Finally, at the 16th degree of polynomial, the model enters the over-parameterized regime. The number of features exceeds the number of training points, but the addition of the roughness penalty in the cost function prevents the model from overfitting excessively:

$$f_\mathbf{w}(x) = w_0 + w_1 x + w_2 x^2 + \cdots + w_{16} x^{16}.$$

This time, the fitted model achieves perfect interpolation of the training points while maintaining smoothness, as dictated by the roughness penalty. This highlights the modern generalization phenomenon, where over-parameterized models can achieve both low training error and stable test performance.

The roughness penalty thus plays a pivotal role in stabilizing the model in the over-parameterized regime, ensuring that adding more polynomial terms does not degrade generalization performance. This contrasts with classical generalization theory, where over-parameterized models are expected to overfit drastically.

4.2.6 Sample non-monotonicity in generalization performance

In both classical and modern generalization theories, there is a prevailing assumption that larger training datasets generally lead to better test performance. However, this assumption does not always hold. In some cases, increasing the size of the training set can lead to worse generalization performance on the test set. This interesting phenomenon, referred to as sample non-monotonicity, highlights the nuanced relationship between the size of the training data and the model's test performance.

To investigate this phenomenon, we repeat the experiment from earlier sections, this time varying the number of training samples. We measure MSE in both the training and test sets for different polynomial degrees and plot the results for multiple training sample sizes. The following code snippet plots multiple MSE curves against the polynomial degree, with each curve corresponding to a different number of training samples.

```python
# Initialize the plot
fig = plt.figure()
ax = fig.add_subplot(1, 1, 1)
ax.set_yscale('log')
ax.set_xlabel('Polynomial degree', labelpad=10)
ax.set_ylabel('MSE', labelpad=10)

# Define the range of training sample sizes
num_train_samples_min = num_train_samples - 4
num_train_samples_max = num_train_samples

# Loop through different training sample sizes
for tmp_num_train_samples in range(num_train_samples_min,
    num_train_samples_max + 1, 2):
    # Compute performance metrics for the current sample size
    mse_train, mse_test = compute_performance(num_train_samples=
        tmp_num_train_samples,
                                    num_runs=num_runs,
                                    num_basis_max=
                                        num_basis_max,
                                    train_noise_std=0)
    # Set color intensity based on the sample size
    e = float(tmp_num_train_samples - num_train_samples_min) /
        float(num_train_samples_max - num_train_samples_min)
    e = 0.2 + 0.6 * e

    # Plot training MSE curve
    ax.plot(np.arange(1, num_basis_max + 1), mse_train, color=(e,
        e, 1.0),
            label=f'Train N={tmp_num_train_samples}', linestyle='
                --')

    # Plot test MSE curve
    ax.plot(np.arange(1, num_basis_max + 1), mse_test, color
        =(1.0, e, e),
            label=f'Test N={tmp_num_train_samples}')

# Add legend and vertical line for interpolation threshold
ax.legend(frameon=False)
ax.axvline(x=9, color='gray', linewidth=0.5, linestyle='--')

# Show the plot
plt.show()
```

Listing 4.8
Assessing Sample Non-monotonicity.

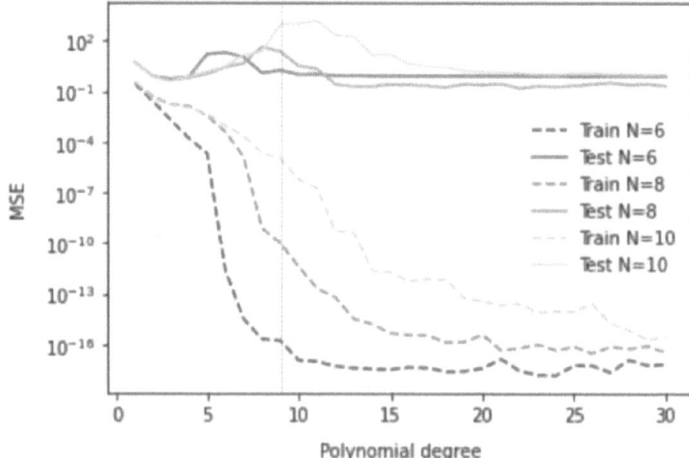

FIGURE 4.9
Observing the sample non-monotonicity phenomenon.

Here, the training sample size is varied to allow us to explore the impact of different sample sizes. The curves for each sample size are also color-coded to better differentiate, with the intensity determined by the relative size of the training set. For each sample size, separate curves are plotted for the training MSE and the test MSE.

The resulting plot, shown in Figure 4.9, reveals a counterintuitive trend: in regions where the polynomial degree is up to the 9th, increasing the size of the training set leads to a higher test MSE. This result highlights the sample non-monotonicity phenomenon and challenges the traditional assumption that more data always improves generalization performance.

Therefore, by analyzing test MSE across varying polynomial degrees and training sample sizes, we observe regions where increasing the training set leads to worse generalization. This highlights the nuanced interplay between training data, model complexity, and generalization performance, motivating further investigation into the conditions under which sample non-monotonicity occurs.

4.3 Summary

The double-descent curve extends the classical U-shaped risk curve observed in the bias-variance tradeoff. In classical settings, increasing model complexity initially reduces bias but increases variance, resulting in a U-shaped test error as a function of model complexity. However, the double-descent curve introduces a second descent in the over-parameterized regime, where the number of parameters far exceeds the number of training samples.

The double-descent curve manifests not only as a function of model complexity but also as a function of training iterations. Early iterations may lead to underfitting, while excessive iterations in over-parameterized models often lead to better test performance due to implicit regularization effects.

Overfitting, characterized by high variance and poor generalization, can be mitigated by increasing the number of parameters to move into the over-parameterized regime. This approach leverages implicit regularization via optimization methods like SGD. Alternatively, we can reduce model capacity through explicit regularization techniques, such as adding penalty terms to the cost function.

Explicit regularization typically involves modifying the cost function. In the smoothing spline framework, the penalty term measures the roughness of the fitted function using its second derivative, which serves as a natural measure of roughness. By penalizing the integrated squared second derivative in the cost function, smoothing splines achieve a balance between goodness of fit and smoothness. This term ensures that the fitted curve avoids excessive oscillations while closely following the training data.

The EMC is a recently proposed concept to quantify the capacity of a model during training. It is related to the maximum number of training samples for which the procedure achieves close to zero training error on average. This metric better reflects the true capacity of a model compared to classical notions of complexity.

Contrary to the conventional belief that more training data always improves performance, the phenomenon of sample non-monotonicity demonstrates that larger training datasets can sometimes worsen test performance. This counterintuitive result depends on the interaction between data distribution, training dynamics, and model complexity.

5

Fundamentals of Deep Neural Networks

Thus far, we have focused on understanding how the addition of features and the control of model complexity influence a model's ability to approximate the true underlying function. For instance, in the preceding chapter, we observed that the approximation capacity of a polynomial function improves with the inclusion of higher-degree terms. This process is essentially a form of functional approximation, where a complex function is represented as a combination of simpler basis functions. The objective, therefore, is to construct a machine-learning model that is sufficiently expressive to closely approximate the true underlying function f.

The addition of parameters introduces a trade-off: while more parameters can enhance the model's expressiveness, they may also lead to over-parameterization. In the classical learning paradigm, over-parameterization often tends to be avoided as it risks overfitting. However, in the modern interpolation regime, this notion has shifted, and over-parameterization is often embraced for its surprising generalization capabilities, which occur in many contexts when the model is properly trained with a certain form of regularization, either implicit or explicit. This paradigm is particularly intriguing in the context of neural networks, where the question arises: What class of functions can be effectively modeled and approximated by these architectures?

The universal approximation theorem for neural networks provides a starting answer to this question. This theorem says that, neural networks possess the theoretical capacity to approximate any continuous function f in a compact domain to an arbitrary degree of accuracy, given a sufficiently powerful width and appropriately chosen weights for the network. This theorem underscores the remarkable flexibility and expressiveness of neural networks, enabling them to map inputs to outputs with virtually any desired level of precision. The crux of this capability lies in the proper training of weights $\mathbf{w}$ and selection of architectural configurations.

In this chapter, we will dive deeper into how neural networks carry out this approximation exercise. We will begin with the fundamental building block of these models: the perceptron. A perceptron is a unit that performs a linear transformation followed by a nonlinear activation. It forms the basis for more complex computations within a network. By stacking multiple perceptrons into layers, neural networks can develop the capacity to approximate highly nonlinear functions, thus justifying their utility as universal approximators.

DOI: 10.1201/9781003511601-5

5.1   Multilayer Perceptron

A multilayer perceptron (MLP) is one of the most fundamental structures in neural networks. It consists of an input layer that represents the input features from the observations X, an output layer that generates predictions $\hat{y}$, and one or more hidden layers in between, representing the black-box transformations that map an input to an output. Each hidden layer comprises one or more neurons, which act as learns that extract hidden features at different levels. Note that the depth of a neural network is defined as the number of hidden layers plus the output layer, excluding the input layer. A perceptron is just a simple neural network that has no hidden layers and consists only of input and output layers.

More specifically, a perceptron performs two mathematical operations: a weighted sum and a nonlinear transformation. For an observation $\mathbf{x} \in \mathbb{R}^p$ with p features, the perceptron first calculates the weighted sum of the input features with their associated weights $\mathbf{w} \in \mathbb{R}^p$ as:

$$z = \sum_{i=1}^{p} w_i x_i = \mathbf{w}^\top \mathbf{x}.$$

To further enhance flexibility, an additional bias term $b \in \mathbb{R}$ is often introduced by default, allowing the perceptron to adjust the weighted sum by a constant shift. This bias term enables the perceptron to better fit the data by introducing a global adjustment to the function's output. Incorporating the bias term, the weighted sum becomes:

$$z = \mathbf{w}^\top \mathbf{x} + b.$$

After computing the weighted sum, the perceptron applies an activation function $\phi(z)$ to introduce nonlinearity to the weighted sum. This step is crucial because, without nonlinear transformations, even a deep neural network would simply reduce to a composition of nested linear functions, which essentially becomes one single linear model. Nowadays, the most popular activation function is called the Rectified Linear Unit (ReLU), defined as:

$$\phi(z) = \max(0, z),$$

which outputs z if $z > 0$ and zero otherwise. ReLU allows the perceptron to "activate", or "fire," for positive inputs while zeroing out, or muting, negative ones. This nonlinearity, combined with potentially complex architecture design (e.g., the number of layers and neurons per layer), provides neural networks with sufficient expressivity and approximation power. Figure 5.1 summarizes the process flow of a perceptron.

Now consider a simple example of a two-layer neural network with a two-dimensional input $\mathbf{x} \in \mathbb{R}^2$. We define both the hidden layer and the output

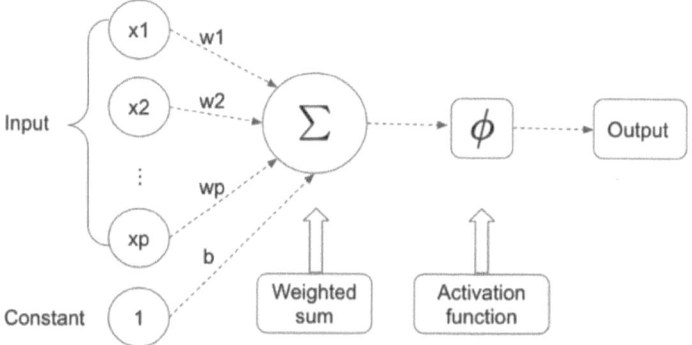

FIGURE 5.1

The process flowchart of a perceptron, which consists of a weighted sum operation followed by an activation function. A column of ones is automatically added to introduce a bias term in the weight vector.

layer to contain only one neuron, respectively. Without activation, the computation in the hidden layer is given by:

$$z = w_{11}x_1 + w_{12}x_2 + b_1,$$

where w_1^1 and w_1^2 are the weights in the first hidden layer, where the subscript denotes the layer index and the superscript denotes the node index. b_1 is the bias term for the first hidden layer. The output of the network, if not using any nonlinear transformation, is then:

$$y = w_2z + b_2 = w_2(w_1^1x_1 + w_1^2x_2 + b_1) + b_2.$$

Expanding this expression yields:

$$y = w_2w_1^1x_1 + w_2w_1^2x_2 + w_2b_1 + b_2,$$

which is clearly a linear function of x_1 and x_2. This linearity also persists even in deeper or wider networks if no activation function is applied. Hence, the introduction of nonlinearity through activation functions like ReLU is essential for us to escape away from linearity and develop more capacity to model complex, nonlinear relationships.

More formally, the operation embedded in a perceptron can be expressed as:

$$f(\mathbf{x}) = \phi(\mathbf{w}^\top\mathbf{x} + b),$$

where ϕ is the activation function, $\mathbf{w}$ is the weight vector, and b is the bias. ReLU ensures that negative values are set to zero, while positive values are retained.

In a multilayer neural network, the operation in an arbitrary l-th layer can be recursively defined as:

$$f^{(l)}(\mathbf{x}) = \phi(\mathbf{W}_l f^{(l-1)}(\mathbf{x}) + \mathbf{b}_l),$$

where $\mathbf{W}_l$ and $\mathbf{b}_l$ are the weight matrix and bias vector for the l-th layer (assuming there are more than one node in a layer), and $f^{(l-1)}(\mathbf{x})$ is the output from the $(l-1)$-th layer. For the first layer, the input is simply $f^{(0)}(\mathbf{x}) = \mathbf{x}$. The final output of the network, with L layers, is expressed as:

$$f^{(L)}(\mathbf{x}) = \phi\left(\mathbf{W}_L \phi\left(\mathbf{W}_{L-1}\phi\left(\cdots\phi\left(\mathbf{W}_1\mathbf{x}+\mathbf{b}_1\right)\cdots\right)+\mathbf{b}_{L-1}\right)+\mathbf{b}_L\right).$$

We can see that it is this repeated application of linear transformations and nonlinearities that enable the network to approximate highly complex functions.

Overall, the computational efficiency of ReLU, particularly during back-propagation, is a key reason for its popularity. When $z \leq 0$, the gradient of ReLU is zero, thus ignoring the parameter update via gradient descent. For $z > 0$, the gradient of ReLU is 1, thus propagating as-is. These properties make ReLU both efficient and effective for training deep networks.

The following code snippet shows how to define the ReLU activation function and apply it to each element in an input list. We see that the value of -1 is set to zero due to the zeroing effect for negative values, while the other two remain the same.

```
def relu(x):
  return np.maximum(x, 0)
>>> relu([1,-1,2])
# Output
array([1, 0, 2])
```

Listing 5.1
Defining ReLU Operation.

Figure 5.2 illustrates the ReLU activation function and its role in introducing nonlinearity. In practice, input $\mathbf{x}$ can also be represented as a tensor in frameworks like PyTorch, facilitating efficient computation for high-dimensional data. This combination of nonlinear activation and architectural flexibility enables neural networks to automatically extract meaningful hierarchical features, such as edges or shapes of objects in image classification tasks, which would be challenging to design and extract manually or using other models.

5.1.1 A two-layer neural network

A two-layer neural network consists of nodes (representing features) and edges (representing weights). The architecture of such a network needs to be manually specified in advance, after which an optimization process is invoked to

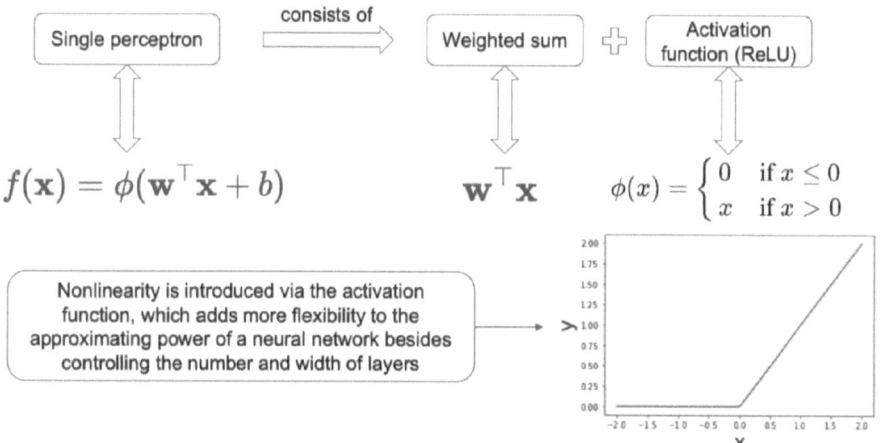

FIGURE 5.2
Decomposing a single perceptron into a weighted sum and an activation function of ReLU.

compute an optimal set of weights based on the given training data and loss function. Multiple factors influence the final out-of-sample performance of the trained model, including the quality of the training data, the suitability of the network architecture, and the quality of the estimated weights. These factors jointly determine the generalization performance of the model.

Let us analyze the approximation capacity of a simple two-layer neural network with a ReLU activation function. This network comprises one hidden layer and one output layer. For simplicity, we assume a one-dimensional scalar input x, a hidden layer with five nodes, and a one-dimensional output. The network can then be expressed as:

$$f(x) = \mathbf{w}_2^\top \phi(\mathbf{w}_1 x + \mathbf{b}_1),$$

where $\mathbf{w}_1 = \{w_1^1, w_1^2, w_1^3, w_1^4, w_1^5\}$, $\mathbf{b}_1 = \{b_1^1, b_1^2, b_1^3, b_1^4, b_1^5\}$, and $\mathbf{w}_2 = \{w_2^1, w_2^2, w_2^3, w_2^4, w_2^5\}$ are all 5×1 vectors of weights and biases, respectively. For simplicity, the scalar bias term b_2 between the hidden and output layers is omitted, and the output layer directly computes a weighted sum without applying an activation function. Each element in the bias vector $\mathbf{b}_1$ corresponds to a unique neuron in the hidden layer. Figure 5.3 illustrates the architecture of the two-layer network, where the hidden layer outputs a 5×1 vector of intermediate features, and the output layer generates a scalar output via a weighted sum of these five hidden features.

To implement this network, we begin by initializing the input data and random weights for both layers. The hidden layer computes the weighted sum followed by applying the ReLU function, and the output layer computes the final result via a linear combination of these activated nodes. The following Python code demonstrates this process:

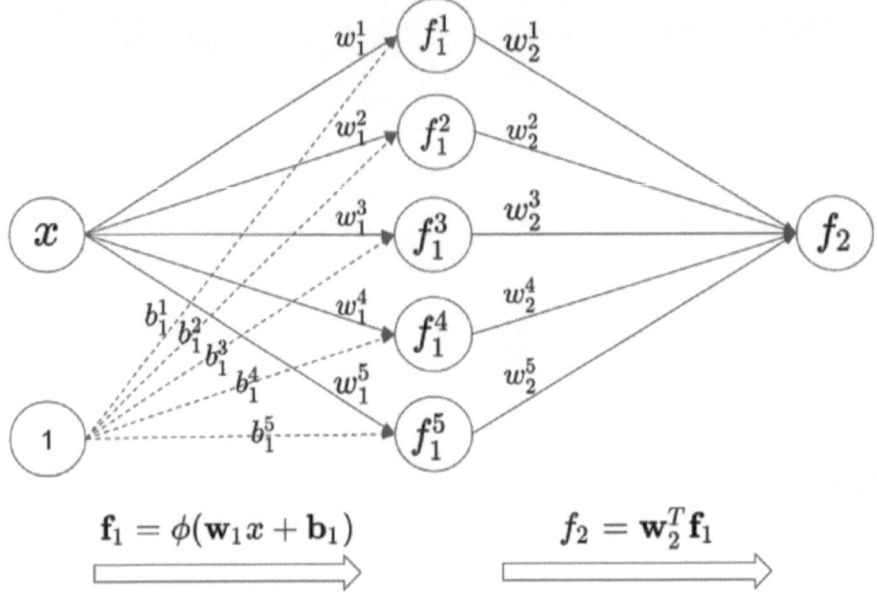

$$\mathbf{f}_1 = \phi(\mathbf{w}_1 x + \mathbf{b}_1) \qquad\qquad f_2 = \mathbf{w}_2^T \mathbf{f}_1$$

FIGURE 5.3
Depicting the architecture of a two-layer network.

```
1   import numpy as np
2   import matplotlib.pyplot as plt
3
4   # Create an array of inputs
5   xs = np.linspace(-2, 2, 100)
6
7   # Randomly initialize the weights and biases
8   W1 = np.random.uniform(low=-1, high=1, size=(5, 1))
9   b1 = np.random.uniform(low=-1, high=1, size=(5, 1))
10  W2 = np.random.uniform(low=-1, high=1, size=(5, 1))
11
12  # Compute hidden layer activations and final output
13  first_layer_activations = np.maximum(0, W1 * xs + b1)
14  final_output = np.dot(W2.T, first_layer_activations)
15
16  # Visualize the activations of individual neurons
17  plt.figure(figsize=(8, 6))
18  plt.plot(np.tile(xs[:, None], 5), first_layer_activations.T)
19  plt.xlabel('$x$')
20  plt.ylabel('Hidden layer activation output')
21  plt.yticks([])
22  plt.show()
```

Listing 5.2
Intermediary Activations.

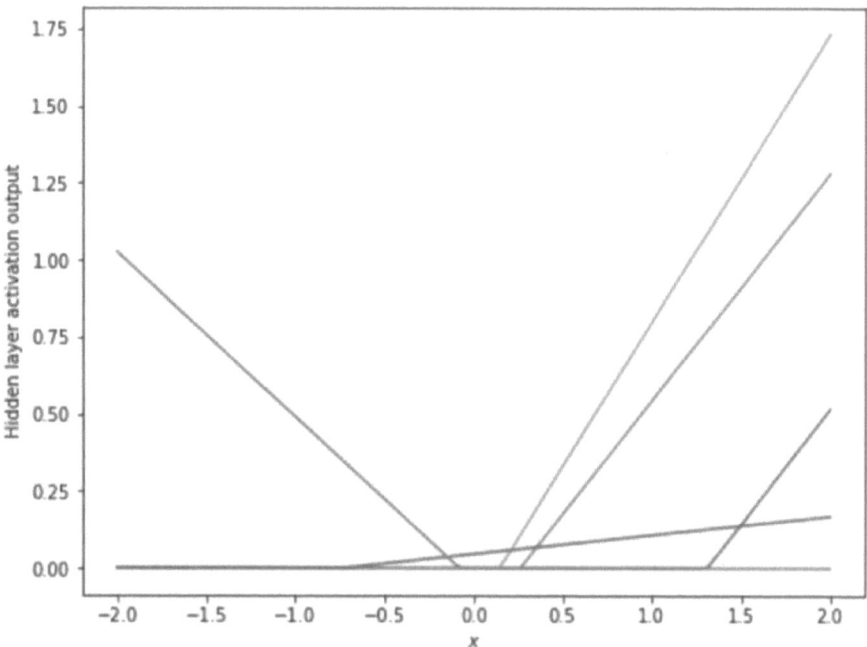

FIGURE 5.4
Visualizing the activation output of the five neurons in the hidden layer.

The code above generates Figure 5.4, showing the activations of the five neurons in the hidden layer. Each activation function has a ReLU-like shape but differs in three aspects: the turning point (threshold), the slope of the linear segment, and the direction of the slope. These activations are derived features (to be) automatically learned in the hidden space, contrasting with the manually engineered polynomial features from the previous chapter. During optimization, these features evolve to better represent the input data, offering diverse perspectives at various levels of granularity. They can also be interpreted as a set of learned basis functions.

The output of the final layer is a linear combination of these basis functions. To further examine the transformations, we visualize the final fitted model and highlight the breakpoints due to these ReLU activations. Breakpoints occur where the ReLU output transitions from zero to positive. These points can be identified by setting the ReLU output to zero and solving for the corresponding input value. The following code achieves this:

```
# Visualize the final output of the two-layer network
plt.figure(figsize=(8, 6))
plt.plot(xs, final_output[0, :], 'k:')   # Plot the final output

# Compute and plot breakpoints
breakpoints = (-b1 / W1).T
```

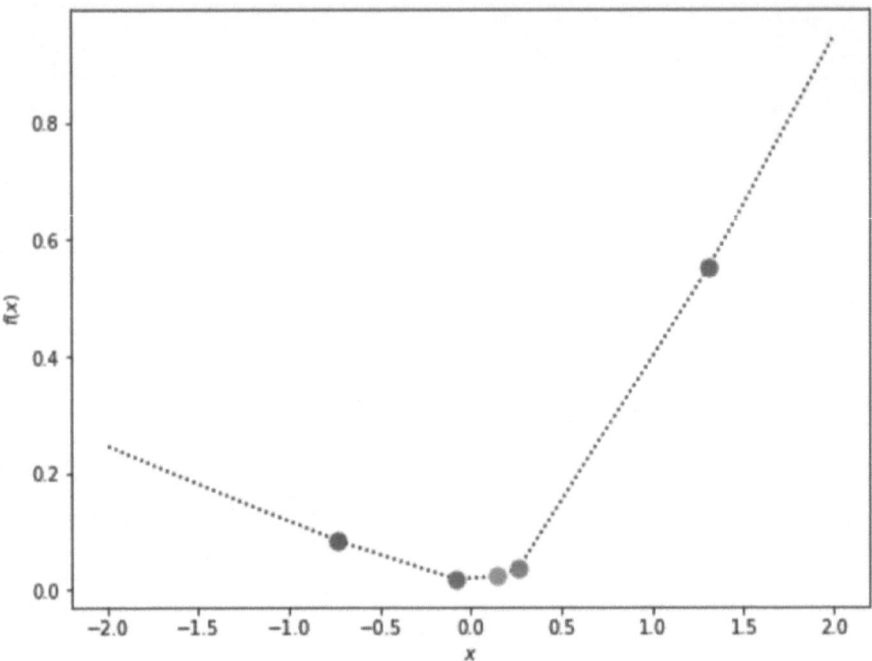

FIGURE 5.5
Visualizing the final layer's output.

```
7   f_breakpoints = np.dot(W2.T, np.maximum(0, W1 * breakpoints + b1)
        )
8   plt.plot(breakpoints, f_breakpoints, '.', markersize=18)
9   plt.xlabel('$x$')
10  plt.ylabel('$f(x)$')
11  plt.show()
```

Listing 5.3
Identifying Breakpoints.

The plot in Figure 5.5 reveals that the final output is piecewise linear, with turning points defined by the intersections of the ReLU activations in the hidden layer. Between these breakpoints, the function behaves linearly, highlighting the piecewise linear nature of the model. The number of breakpoints is determined by the number of activation functions in the hidden layer, meaning that the complexity of the final output is linearly related to the number of neurons in the hidden layer.

The expressivity of a neural network is directly influenced by the width of its hidden layers. Increasing the number of nodes in a shallow neural network enhances its ability to model complex functions but comes with an increased computational cost. Instead of widening the network, a more efficient approach is to increase its depth by stacking additional layers, forming a deep neural network. Models with multiple hidden layers, often referred to as deep neural networks, achieve greater approximating power with fewer computational resources, making them the preferred choice for large-scale problems.

5.1.2 Shallow versus deep neural networks

We know that a shallow neural network, which could be composed of one hidden layer followed by an output layer, offers limited flexibility. The model's expressivity is constrained by the number of nodes (or neurons) in the hidden layer. If we want an additional breakpoint or linear segment in the output, we need to add another node in the hidden layer. These hidden neurons act as basis functions that are linearly combined to generate the final output. Consequently, the complexity of a shallow neural network grows linearly with the number of hidden nodes.

In contrast, a deep neural network achieves significantly greater expressivity with fewer nodes per layer, where each layer is stacked on top of the other. Specifically, the complexity of a deep neural network, when measured by the number of breakpoints or linear segments in its output, grows exponentially with the number of hidden layers. To illustrate, we will consider a hand-designed deep neural network using ReLU activation functions and analyze how its complexity evolves as the network becomes deeper.

Suppose that we wish to implement the following "sawtooth" network model, where the relationship between the current layer $f_l(x)$ and the previous layer $f_{l-1}(x)$ is given by:

$$f_l(x) = 2|f_{l-1}(x)| - 2.$$

For the initial layer, we set:

$$f_0(x) = x.$$

The absolute value function can be decomposed into two cases: if the input is positive, the output remains unchanged; if the input is negative, the output is negated. Using ReLU, the absolute value function can be implemented as:

$$f_l(x) = 2\text{ReLU}(f_{l-1}(x)) + 2\text{ReLU}(-f_{l-1}(x)) - 2.$$

Here, the weights are fixed, and no learning occurs. The focus is on the approximating power of this structured network. Using the code snippet below, we define a two-neuron hidden layer that implements this transformation:

```
# Define a single layer of the sawtooth network
def layer(x):
    return 2 * np.maximum(0, x) + 2 * np.maximum(0, -x) - 2

Next, we build the full network by sequentially stacking these
    layers. The number of layers can be specified to control the
    depth of the network:

# Define a multilayer sawtooth network
```

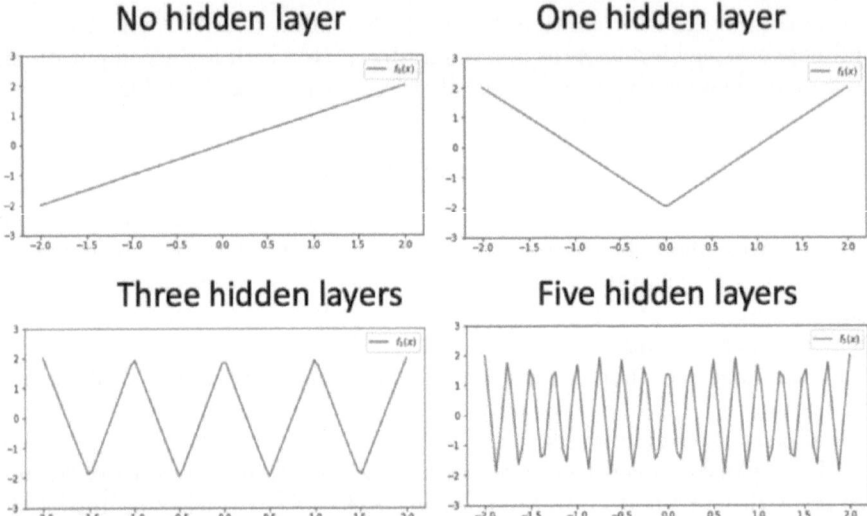

FIGURE 5.6

Visualizing model output with varying numbers of hidden layers.

```
 8  def multilayer(x, hidden_layers=1):
 9      f = x  # First layer output is the input itself
10      for _ in range(hidden_layers):  # Sequentially stack layers
11          f = layer(f)
12      return f
```

Listing 5.4

Implementing Sawtooth Network.

By varying the number of hidden layers, we can observe how the output of the network changes. For example, we can compute and visualize the output for networks with zero to five hidden layers:

```
1  # Visualize the outputs for varying numbers of hidden layers
2  numlayers = 6
3  fig, axs = plt.subplots(numlayers, 1, figsize=(8, 25))
4  for i, ax in enumerate(axs):
5      ax.plot(xs, multilayer(xs, hidden_layers=i))
6      ax.set_ylim([-3, 3])
7      ax.legend([f'$f_{{{i}}}(x)$'])
8  plt.show()
```

Listing 5.5

Varying the Number of Hidden Layers.

The outputs reveal exponential growth in complexity as we increase the number of hidden layers, as shown in Figure 5.6.

We analyze four cases in the following:

- No hidden layer: The output is simply the input itself, as shown in the top left panel of Figure 5.6:

$$f_0(x) = x.$$

- One hidden layer: The output consists of two linear segments, corresponding to the absolute value function implemented by two ReLU activations, as shown in the top right panel of Figure 5.6:

$$f_1(x) = 2|x| - 2.$$

- Three hidden layers: With three hidden layers, the output consists of $2^3 = 8$ linear segments, as illustrated in the bottom left panel of Figure 5.6. This exponential growth demonstrates the increasing approximation power of deeper networks, where the final output can be expressed as:

$$f_3(x) = 2\text{ReLU}(f_2(x)) + 2\text{ReLU}(-f_2(x)) - 2.$$

- Five hidden layers: With five hidden layers, the number of breakpoints grows to $2^5 = 32$, as shown in the bottom right panel of Figure 5.6. In comparison, a shallow network with 10 nodes in one hidden layer could only achieve a maximum of 10 breakpoints, highlighting the advantage of depth in increasing expressivity.

Deep networks attribute their superior representational power to this exponential growth in breakpoints and linear segments. Despite using the same total number of neurons, deep neural networks can approximate far more complex functions than shallow ones.

Another key advantage of deep networks is their ability to learn hierarchical features. Shallow networks often tend to "memorize" training data, leading to poor generalization. In contrast, deep networks learn features at multiple levels of abstraction. For example, in image recognition tasks, early hidden layers tend to capture low-level features such as edges and corners of common objects, while later layers focus on higher-level patterns like shapes and objects.

In summary, the number of breakpoints and linear segments acts as indicators of model complexity, which increases exponentially with the depth of a neural network. This exponential growth in expressivity makes deep networks powerful enough to approximate functions of arbitrary complexity. Consequently, making the network deeper often provides a more efficient way to increase model capacity compared to simply widening the network.

Next, we will explore automatic differentiation, a mechanism embedded in modern deep learning frameworks that enables fast and efficient gradient computations, facilitating the training of large neural networks.

5.2   Automatic Differentiation

The strong approximation capacity of a deep neural network makes it a highly expressive and powerful model. However, achieving this capacity requires a proper optimization procedure to identify the optimal weights for a given network architecture. In Chapter 3, we explored the family of gradient descent methods, which serve as the backbone of neural network optimization. This general procedure involves iteratively computing the gradient of the model's cost function with respect to the model's weights and subsequently applying this gradient to update the weights.

Gradient-based optimization is one of the primary reasons for the success of neural networks. Specifically, neural networks trained using stochastic gradient descent (SGD) algorithms often achieve low empirical risk and generalize effectively to unseen test data. A distinguishing feature of SGD lies in its stochasticity: the gradient is computed based on a single or a mini-batch of examples, resulting in a noisy gradient compared to the full-batch gradient derived from all training examples.

This stochasticity plays a critical role in enhancing generalization. Unlike full gradient descent, which can drive the optimizer into sharp local minima, SGD with smaller batch sizes facilitates movement toward flatter minima, which tend to generalize better to test data. The noise introduced by stochastic gradients enables the optimizer to escape sharp local optima and explore regions of the parameter space associated with more robust solutions. Consequently, the generalization capacity of neural networks depends not only on the architecture and sample size but also on the specific optimization algorithm and its configuration, such as the batch size and the learning rate.

Consider the trade-off between batch size and noise in gradient calculations. While smaller batch sizes introduce more noise into the optimization process, they are computationally efficient and can potentially improve test performance. On the other hand, larger batch sizes reduce noise but increase the risk of overfitting to sharp local minima. This highlights the intricate interplay between optimization dynamics and model generalization.

We will delve deeper into regularization techniques from the perspective of optimization in a later chapter of the book series. For now, let us examine the intricacies of gradient-based optimization, particularly in the context of neural networks with many parameters that require simultaneous updates during each iteration. The modern way to handling these simultaneous updates is via automatic differentiation, which is a computational mechanism embedded in many modern deep learning frameworks. Automatic differentiation efficiently computes gradients for arbitrary computational graphs, enabling the training of complex models with minimal manual effort. By automating the differentiation process, frameworks like PyTorch and TensorFlow reduce the complexity

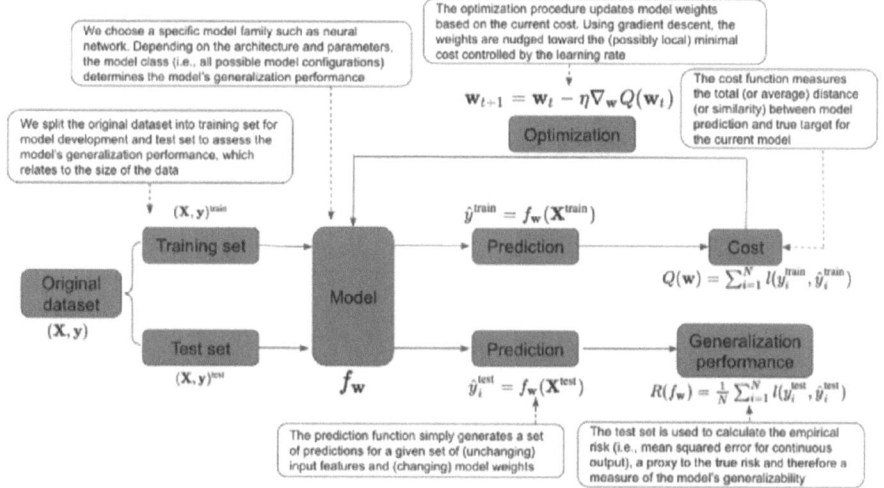

FIGURE 5.7
The model training process.

of deriving and implementing gradient calculations, making deep learning both accessible and scalable.

In the next section, we will explore the mechanics of automatic differentiation and its implementation in neural network optimization.

5.2.1 Gradient-based optimization

Let us revisit the model training process to better understand the role of gradient-based optimization. As shown in Figure 5.7, the dataset is first divided into a training set and a test set. The training set is used to develop an optimal model, while the test set evaluates the generalization performance of the trained model. Two critical functions underpin this process: the prediction function and the cost function. The prediction function, denoted as $f(\mathbf{x}^{(i)}; \mathbf{w})$, takes fixed input features $\mathbf{x}^{(i)}$ and adjustable model weights $\mathbf{w}$ to generate predictions. The cost function then aggregates these predictions and compares them with the true targets to obtain a scalar value, $Q(\mathbf{w})$, which represents the discrepancy between predictions and target labels, often as an average over all observations.

This scalar cost is then passed to an optimization procedure, such as gradient descent, to update the weights $\mathbf{w}$ in a way that minimizes the cost. Once training concludes, the optimal weights $\mathbf{w}^*$ derived from the training set are applied to the test set to generate test set predictions. Comparing these predictions with the ground truth gives us the final cost, such as mean squared error (MSE). This test set cost serves as an approximation to the unknown true risk and reflects how well the model generalizes to unseen data. The process then repeats over multiple epochs, iteratively minimizing the empirical risk until a given stopping criterion is met.

Note that the optimization process hinges on updating weights based on the gradient of the cost function with respect to the current weights. Specifically, gradient descent requires the computation of the gradient vector $\nabla_{\mathbf{w}} Q(\mathbf{w}_t)$, containing the partial derivatives of the cost function Q with respect to each component of $\mathbf{w}$ at iteration t. For neural networks, $\mathbf{w}$ includes all weights across layers.

As discussed earlier, the MLP can be expressed as a nested composite function:

$$f_L(\mathbf{x}) = \phi(\mathbf{b}_L + \mathbf{W}_L \phi(\mathbf{b}_{L-1} + \mathbf{W}_{L-1} \phi(\cdots \phi(\mathbf{b}_1 + \mathbf{W}_1 \mathbf{x}) \cdots))),$$

where ϕ represents the activation function, $\mathbf{W}_l$ the weight matrix, and $\mathbf{b}_l$ the bias vector at layer l. The final prediction $f_L(\mathbf{x})$ thus depends on all intermediate weights, including those in the first layer, $\mathbf{W}_1$ and $\mathbf{b}_1$. To simplify, let $\mathbf{w}$ denote all network weights, such that the prediction becomes a function of these weights: $f_L(f_{L-1}(\cdots f_1(\mathbf{w}) \cdots))$. For each l-th layer out of L layers, the function f_l is a differentiable mapping from $\mathbb{R}^{p_{l-1}}$ to $\mathbb{R}^{p_l}$.

Differentiability of f_l is crucial for gradient-based optimization. The gradient vector $\nabla_{\mathbf{w}} Q(\mathbf{w})$ is computed using the chain rule to propagate gradients through the layers, enabling updates to the weights $\mathbf{w}$. If any intermediate function f_l is non-differentiable, gradient-based methods cannot proceed, thus halting the optimization process.

5.2.2   The chain rule with partial derivatives

The chain rule, integral to computing partial derivatives for composite functions, underpins gradient calculations in neural networks. Consider a composite function $f(g(x))$, where f operates on the output of g. To differentiate $f(g(x))$, the chain rule says:

$$\frac{d}{dx}[f(g(x))] = f'(g(x))g'(x).$$

This calculation begins with the derivative of the outer function f with respect to the inner function $g(x)$, yielding $f'(g(x))$. The derivative of $g(x)$, which is $g'(x)$, is computed next, and the two derivatives are then multiplied. This backward flow continues through all intermediate functions composing the final expression of the nested function.

In neural networks, the MLP structure consists of stacked layers that form a composite function. Using the chain rule, we calculate partial derivatives for the network weights by propagating derivative calculations layer by layer, starting from the final output and traversing backward to the input. This sequence mirrors the inverse of the forward propagation process, where inputs are transformed through matrix multiplications and nonlinearities, yielding

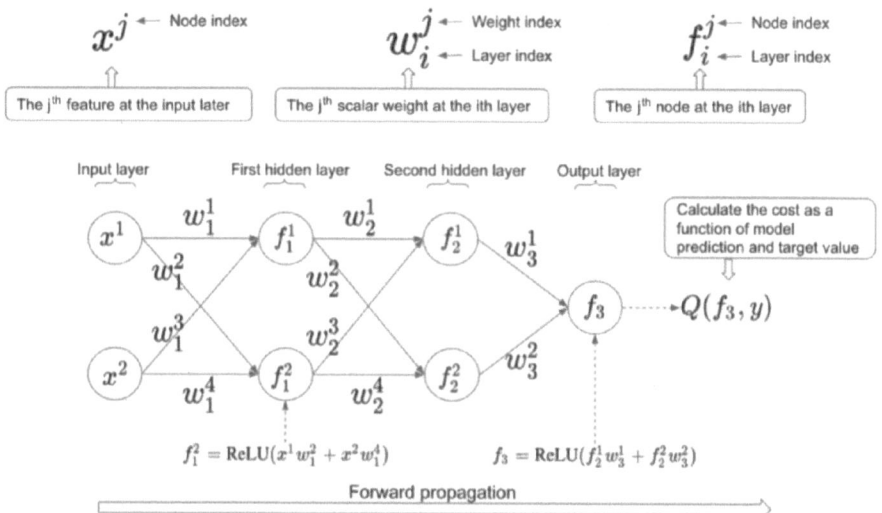

FIGURE 5.8
Schematic of forward propagation in a simple neural network with two hidden layers, each having two nodes.

predictions. Backpropagation, the backward pass, calculates partial derivatives for each weight by propagating the gradient of the cost function backward through the network.

For example, consider a simple neural network with two hidden layers (two nodes each), two input nodes, and one output node, as illustrated in Figure 5.8. Let $\mathbf{x}$ denote input features, $\mathbf{w}$ the weights, and f the function that produces outputs from inputs. The index notation captures the layer structure: x^j is the jth input feature, w_i^j the jth weight in layer i, and f_i^j the jth node output in layer i.

The forward propagation process involves repeatedly applying matrix multiplications and activation functions. For example, in the first hidden layer, the second node computes $f_1^2 = \text{ReLU}(x^1 w_1^2 + x^2 w_1^4)$. This learned feature becomes an input to the next layer. In the final layer, the model predicts $f_3 = \text{ReLU}(f_2^1 w_3^1 + f_2^2 w_3^2)$, and the cost $Q(f_3, y)$ is calculated.

Backward propagation begins by computing the gradient of the cost with respect to the weights in the last layer, then iteratively moves backward, calculating gradients for earlier weights using the chain rule. For example, to compute $\frac{\partial Q}{\partial w_1^1}$, one traces the derivative path from Q through all intermediate layers to w_1^1, summing contributions from all paths, as shown by the dash-dotted lines in Figure 5.9.

Let the prediction function for a neural network be represented as $f_L(f_{L-1}(\cdots f_1(\mathbf{w})\cdots))$, where f_L is the output function, and $\mathbf{w}$ denotes the

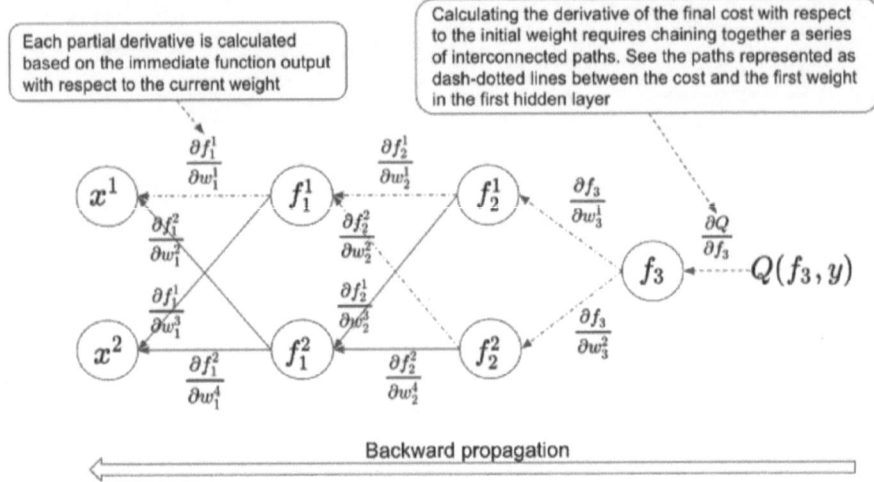

FIGURE 5.9
Illustration of the backward propagation process.

initial weights. To compute $\frac{\partial f_L}{\partial \mathbf{w}}$, we sequentially calculate intermediate derivatives $\frac{\partial f_L}{\partial f_{L-1}}$, $\frac{\partial f_{L-1}}{\partial f_{L-2}}$, down to $\frac{\partial f_1}{\partial \mathbf{w}}$, then multiply them:

$$\frac{\partial f_L}{\partial \mathbf{w}} = \frac{\partial f_L}{\partial f_{L-1}} \cdot \frac{\partial f_{L-1}}{\partial f_{L-2}} \cdots \cdots \frac{\partial f_2}{\partial f_1} \cdot \frac{\partial f_1}{\partial \mathbf{w}}.$$

This process, depicted in Figure 5.10, often relies on automatic differentiation frameworks such as PyTorch. The forward pass generates predictions, while the backward pass computes intermediate derivatives and multiplies them to produce gradients, completing the optimization cycle.

5.2.3 Different modes of multiplication

The process of calculating gradients in multilayer neural networks involves multiplying several matrices of partial derivatives across layers, with each element representing a unique pair of nodes between consecutive layers. Efficient computation of these gradients, especially for large networks, requires careful planning of the multiplication order to minimize computational cost.

Consider the multiplication of the first two matrices: $\frac{\partial f_1}{\partial \mathbf{w}}$ and $\frac{\partial f_2}{\partial f_1}$. Let $\mathbf{w}$ represent a p_0-dimensional vector and f_1 a p_1-dimensional vector. The derivative $\frac{\partial f_1}{\partial \mathbf{w}}$ forms a $p_1 \times p_0$ matrix, while $\frac{\partial f_2}{\partial f_1}$ is a $p_2 \times p_1$ matrix. Matrix multiplication requires that the inner dimensions match, which can then produce a $p_2 \times p_0$ matrix with a computational cost of $p_2 p_1 p_0$ operations. This cost indicates how long the multiplication will take.

Extending this process to the entire network, the sequence of matrix multiplications determines the computational complexity. In forward mode

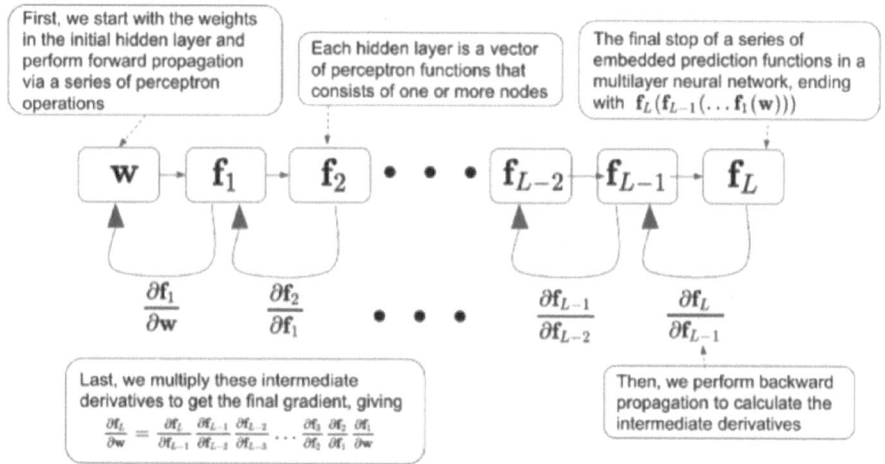

FIGURE 5.10
Calculating the gradient of a multilayer neural network with respect to the weights in the initial hidden layer.

multiplication, the matrices are multiplied sequentially from the input layer to the output layer. For a network with L layers, this sequence can be represented as:

$$\frac{\partial f_L}{\partial \mathbf{w}} = \frac{\partial f_L}{\partial f_{L-1}} \left(\frac{\partial f_{L-1}}{\partial f_{L-2}} \left(\cdots \left(\frac{\partial f_3}{\partial f_2} \left(\frac{\partial f_2}{\partial f_1} \frac{\partial f_1}{\partial \mathbf{w}} \right) \right) \cdots \right) \right).$$

As shown in Figure 5.11, each successive multiplication with an outer layer matrix changes the size of the resulting matrix and incurs an additional calculation workload. For example, $\frac{\partial f_3}{\partial f_2} \left(\frac{\partial f_2}{\partial f_1} \frac{\partial f_1}{\partial \mathbf{w}} \right)$ results in a $p_3 \times p_0$ matrix, incurring $p_3 p_2 p_0$ operations. Summing these across all layers gives the total computational cost for forward-mode multiplication, we have:

$$p_0 \sum_{l=2}^{L} p_l p_{l-1}.$$

On the other hand, in a backward-mode multiplication, matrices are multiplied in reverse order, starting from the output layer and proceeding toward the input layer. The sequence is expressed as:

$$\frac{\partial f_L}{\partial \mathbf{w}} = \left(\cdots \left(\left(\left(\frac{\partial f_L}{\partial f_{L-1}} \frac{\partial f_{L-1}}{\partial f_{L-2}} \right) \frac{\partial f_{L-2}}{\partial f_{L-3}} \right) \cdots \frac{\partial f_2}{\partial f_1} \right) \frac{\partial f_1}{\partial \mathbf{w}} \right).$$

As shown in Figure 5.12, the computational cost for backward-mode multiplication is:

$$p_L \sum_{l=1}^{L-1} p_l p_{l-1}.$$

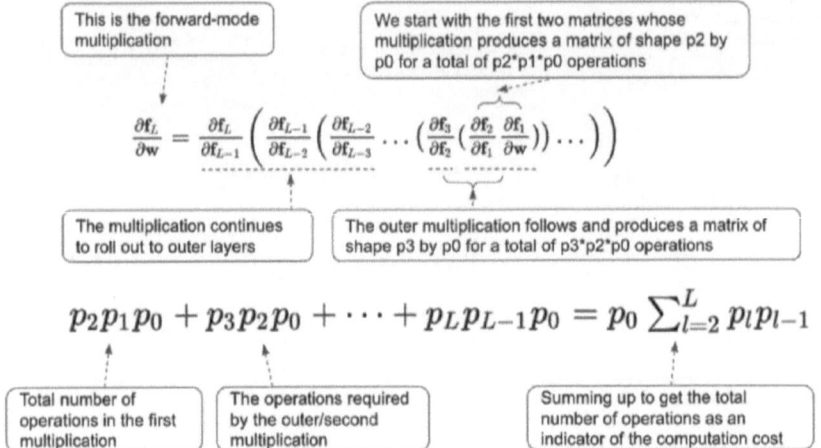

FIGURE 5.11
Illustrating the forward-mode multiplication process and its computational cost.

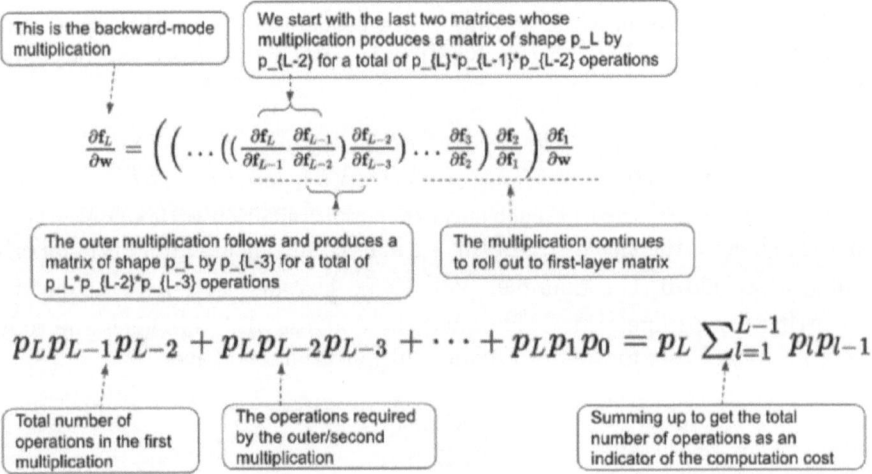

FIGURE 5.12
Illustrating the backward-mode multiplication process and its computational cost.

Comparing these two modes, the key difference is that the forward mode mainly depends on p_0, the dimension of the input data, while the backward mode mainly depends on p_L, the output dimension. Since the cost function in most models is a scalar ($p_L = 1$), backward-mode multiplication is generally more efficient, particularly for high-dimensional inputs like image data, where p_0 can reach hundreds or thousands.

However, despite its computational efficiency, backward-mode multiplication, or backpropagation, requires more memory to store intermediate derivatives. In contrast, the forward mode calculates the layer-wise gradients during the forward pass and discards them afterward, reducing memory usage. This trade-off is critical when designing automatic differentiation frameworks such as PyTorch, which offers efficient implementation of backpropagation when training large neural networks.

In the next section, we will dive into the implementation details of training a simple convolutional neural network (CNN) model using the Modified National Institute of Standards and Technology (MNIST) dataset.

5.3   Training a CNN on MNIST

To understand the training process in practice, we explore the implementation of an image classifier using the MNIST dataset. The MNIST dataset, short for the Modified National Institute of Standards and Technology (MNIST) database, is one of the most widely used datasets for prototyping and benchmarking machine-learning and deep learning models. It contains grayscale images of handwritten digits, each normalized to a standard format. Specifically, it comprises 60,000 training images and 10,000 testing images, each representing a single digit from 0 to 9. These images are preprocessed through normalization and center cropping for consistency, ensuring that variations in input do not adversely affect the training process. Accessing the dataset in PyTorch is straightforward, as it is readily available via the *torchvision.datasets* module, eliminating the need for manual downloading.

PyTorch, a core deep learning framework, provides the torch package, which includes the foundational classes and methods for building and training neural networks. This package handles the implementation of tensor operations, automatic differentiation, and the various utilities required for optimization algorithms. Beyond this core functionality, the *torchvision* package serves as a specialized extension for computer vision tasks. It offers a suite of tools, including access to popular datasets like MNIST, prebuilt model architectures, and standard image transformation pipelines. These utilities greatly simplify and standardize image preprocessing, model initialization, and benchmarking. Meanwhile, torchtext caters to natural language processing (NLP) tasks, offering tools for handling textual data and preprocessing pipelines for tokenization, embedding, and other NLP-specific transformations.

To train a simple convolutional neural network (CNN) using the MNIST dataset, we begin with data loading and preprocessing. PyTorch's datasets and DataLoader classes, available within the *torchvision* package, can be used to handle data efficiently. These utilities allow for consistent handling of data pipelines, from accessing the dataset to applying transformations like normalization and efficiently batching the data during training.

Once the data are prepared, the next step involves defining the model architecture. A CNN model is constructed with layers tailored for extracting spatial and hierarchical features from the images. The design typically includes convolutional layers, pooling layers, and fully connected layers, combined with activation functions such as ReLU to introduce nonlinearity.

After defining the architecture, we specify a cost function and an optimization algorithm. The cost function measures the discrepancy between the predicted labels and the ground-truth labels. Common choices for classification tasks include the cross-entropy loss (CEL), which computes the negative log-likelihood of the true class. The optimization algorithm, such as SGD or Adam, is responsible for adjusting the model's parameters to minimize the cost function, leveraging the gradients computed via backpropagation.

Finally, the training process begins. The model iterates over the training data in batches, performing forward propagation to compute predictions, calculating the cost, and using backward propagation to update the weights. This process repeats for a specified number of epochs, gradually refining the model parameters. Once training is complete, the model is evaluated on the test set, where its predictive performance is measured using predictive accuracy or another suitable metric. This step assesses the model's ability to generalize to unseen data.

The utility of PyTorch and its subpackages, *torchvision* and *torchtext*, lies in their ability to streamline this workflow. By providing prebuilt tools and abstractions for data handling, model design, and optimization, these packages significantly reduce the effort required to implement and train deep learning models, allowing researchers to focus on fine-tuning and innovation rather than implementation details.

The following sections show the full training procedure.

5.3.1   Downloading and loading the MNIST dataset

The process of working with the MNIST dataset in PyTorch begins with downloading the dataset and preparing it for training and testing. PyTorch's *torchvision* library provides the *datasets* submodule, which makes accessing popular datasets like MNIST straightforward. Recall that the dataset consists of grayscale images of handwritten digits, with corresponding labels representing the digit each image depicts. It includes 60,000 training images and 10,000 testing images.

The following code snippet demonstrates how to use the *datasets. MNIST* class to download the dataset into a local directory called *data*. The dataset is transformed into PyTorch tensors using the *ToTensor* function, which converts the image data into a multidimensional tensor. This transformation scales the pixel values from their original range of $[0, 255]$ to $[0.0, 1.0]$ for normalization purposes, making the data more suitable for neural network training.

```
1  import torch
2  import matplotlib.pyplot as plt
3  from torchvision import datasets
4  from torchvision.transforms import ToTensor
5
6  train_data = datasets.MNIST(
7      root='data',
8      train=True,
9      transform=ToTensor(),
10     download=True,
11 )
12
13 test_data = datasets.MNIST(
14     root='data',
15     train=False,
16     transform=ToTensor(),
17 )
```

Listing 5.6
Downloading MNIST Dataset.

Once the dataset is downloaded, we can print the object's *train_data* and *test_data* to examine their attributes. The *datasets.MNIST* object provides metadata such as the number of samples, the root location where the data are stored, and any applied transformations.

```
1  >>> print(train_data)
2  Dataset MNIST
3      Number of datapoints: 60000
4      Root location: data
5      Split: Train
6      StandardTransform
7  Transform: ToTensor()
8
9  >>> print(test_data)
10 Dataset MNIST
11     Number of datapoints: 10000
12     Root location: data
13     Split: Test
14     StandardTransform
15 Transform: ToTensor()
```

Listing 5.7
Examining Data Attributes.

Each image in the training dataset has a size of 28×28 pixels, and the corresponding labels are stored in the *targets* attribute. We can confirm this using the following commands:

```
1  >>> print(train_data.data.size())
2  >>> print(train_data.targets.size())
3  torch.Size([60000, 28, 28])
4  torch.Size([60000])
```

Listing 5.8
Checking Data Size.

The output indicates that the dataset contains 60,000 training images, each represented as a 28 × 28 tensor. For grayscale images, an additional channel dimension (depth) is omitted, as each pixel is a single intensity value.

To better understand the dataset, we can visualize a random selection of 25 images from the training set. Each image is plotted on a 5 × 5 grid.

```
figure = plt.figure(figsize=(10, 8))
cols, rows = 5, 5
# Loop over 25 places to plot the images
for i in range(1, cols * rows + 1):
    # Generate a random index to select an image
    # The item function converts the Tensor object into a scalar
        value
    sample_idx = torch.randint(len(train_data), size=(1,)).item()
    # Extract the image data and target label
    img, label = train_data[sample_idx]
    figure.add_subplot(rows, cols, i)
    plt.title(label)
    plt.axis("off")
    # Squeeze the image to convert the image shape from [1,28,28]
        to [28,28]
    plt.imshow(img.squeeze(), cmap="gray")
plt.show()
```

Listing 5.9
Visualizing MNIST Dataset.

The resulting graph shows 25 randomly selected images, where each image consists of 28 × 28 = 784 pixels. These pixels form the features of the image that will be used to train the model. The output of this visualization is shown in Figure 5.13.

The next step involves preparing the dataset for training using PyTorch's *DataLoader* class, which facilitates efficient data handling. The *DataLoader* takes care of batching, shuffling, and parallel processing of the dataset, enabling streamlined training workflows.

```
from torch.utils.data import DataLoader

loaders = {
    'train': torch.utils.data.DataLoader(
        # data source to be loaded
        train_data,
        # the number of training samples used in one iteration
        batch_size=100,
        # samples are shuffled and loaded in batches
        shuffle=True
    ),
    'test': torch.utils.data.DataLoader(
        test_data,
        batch_size=100,
        shuffle=True
    )
}
```

Listing 5.10
Defining Dataloader.

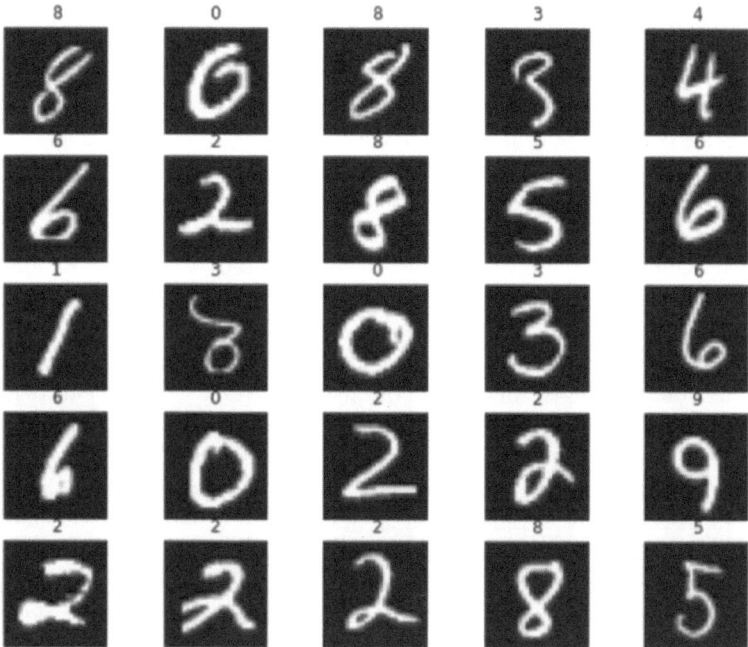

FIGURE 5.13
Visualizing 25 random MNIST digits. Each digit is a grayscale image that consists of $28 \times 28 = 784$ pixels.

Here, the training and test datasets are loaded in batches of size 100, with the training data shuffled to ensure randomness during each epoch. The loaded data can be inspected by iterating over the *DataLoader*.

```
>>> for X, y in loaders['train']:
>>>     print("Shape of X [batch_size, channel, height, width]: ",
    X.shape)
>>>     print("Shape of y: ", y.shape)
>>>     break
Shape of X [batch_size, channel, height, width]:    torch.Size
    ([100, 1, 28, 28])
Shape of y:   torch.Size([100])
```

Listing 5.11
Inspecting Data.

The output confirms that the images are batched with dimensions $[100, 1, 28, 28]$, where 100 is the batch size, 1 represents the grayscale channel, and 28×28 is the spatial resolution of each image. The labels are stored in a tensor of size $[100]$ corresponding to the 100 images in each batch.

This data preparation pipeline ensures that the dataset is ready for training, allowing us to focus on defining the model architecture, loss function, and optimization procedure in subsequent steps.

5.3.2   Defining the prediction function

In machine learning, the prediction function serves as the computational back-bone of any model. It takes input data, processes it through a series of operations involving model parameters (e.g., weights and biases), and outputs predictions. These operations typically involve matrix-vector multiplication to extract features from input data and apply nonlinear transformations such as activation functions to capture complex patterns. A well-defined prediction function achieves two essential goals: initializing model parameters and specifying the architecture of the model, that is, how the data interact with the parameters.

In simpler machine-learning problems, these two operations can often be combined into a single function. For instance, in a basic linear regression model, the prediction function not only defines the weights but also specifies the interaction between the weights and the input data through a dot product. However, as the complexity of the problem grows—whether due to larger datasets, more sophisticated architectures, or advanced optimization techniques—modularizing the prediction function becomes critical. By separating the definition of model parameters from the architectural blueprint, we achieve greater flexibility, scalability, and maintainability in the codebase.

A common approach in deep learning frameworks, such as PyTorch, is to encapsulate the entire model architecture and its parameters in a class. This class serves as a blueprint for creating instances of the model, with two primary components:

- *__init__* Function: This is the constructor of the class, responsible for defining and initializing the essential building blocks of the model. These components include convolutional layers, activation functions, pooling layers, and fully connected layers. In neural networks, these elements correspond to trainable parameters (e.g., weights and biases) that will be optimized during training. Think of this step as assembling the "raw materials" or "parts" needed to build the model.

- *forward* Function: This function defines the flow of data through the network. It specifies how the input interacts with the initialized components, such as passing through convolutional layers, applying activation functions, and eventually producing the output. The forward function acts as an instruction manual for assembling the model's components into a computational pipeline. In PyTorch, when an instance of the class is called, the forward function is automatically invoked, simplifying the process of defining the data flow.

This separation of initialization and data flow provides better clarity for our understanding of the overall training procedure. For example, in PyTorch, the model can be instantiated and immediately called with input data, triggering the forward pass implicitly. This design pattern promotes modularity and aligns with object-oriented programming principles.

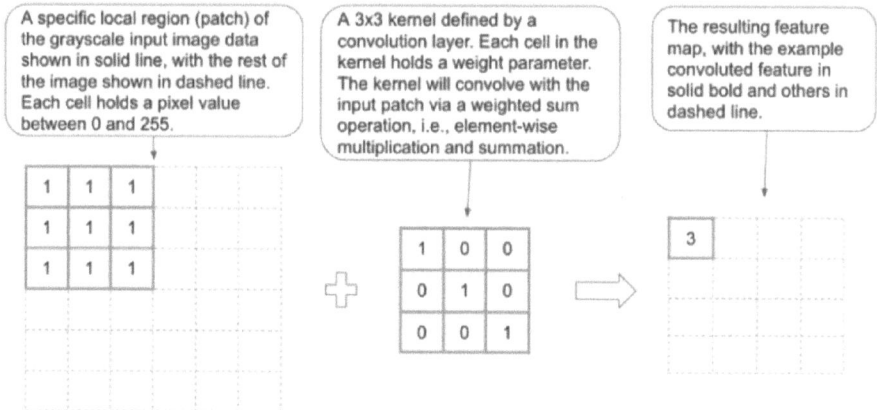

FIGURE 5.14

Illustrating the convolution workflow. Here we define a 3×3 kernel that holds a total of 9 weights. The kernel will convolve with different patches of the same size in the input image data.

CNNs are particularly suited for processing image data because they exploit spatial hierarchies. Unlike fully connected layers, which treat all input features equally, convolutional layers focus on local patterns by applying small, learnable filters (or kernels) to regions of the input. Each kernel slides across the input, performing an element-wise multiplication with the corresponding patch of the input matrix, followed by summation. The result of this operation is a single scalar value that forms part of the feature map.

Figure 5.14 illustrates this process. Consider a 3×3 kernel interacting with a 3×3 patch of an image. Each cell in the patch contains a pixel value (e.g., in the range $[0, 255]$ for grayscale images). The kernel values, which are the trainable parameters of the layer, are multiplied element-wise with the pixel values, and the products are summed to produce a single scalar output. The kernel then moves or slides to the adjacent window. This process is repeated across the entire input image, effectively "scanning" the input to detect patterns such as edges or textures.

A convolution operation can be characterized by the following properties:

- Kernel Size: Determines the dimensions of the filter. For example, a 5×5 kernel captures broader patterns than a 3×3 kernel.

- Stride: Specifies the step size for sliding the kernel. A stride of 1 means the kernel moves one pixel at a time, whereas a stride of 2 skips every other pixel, reducing the spatial dimensions of the resulting output.

- Padding: Adds zeros around the input image to preserve its dimensions after convolution, ensuring that edge information is not lost.

In addition to convolution, CNNs often include activation functions (e.g., ReLU) to introduce nonlinearity and pooling layers to reduce the spatial dimensions of feature maps. Pooling simplifies the representation, focusing on the most salient features. For example, max pooling selects the maximum value in a region, while average pooling computes the mean.

This CNN architecture can be encapsulated in a PyTorch class as follows:

```python
import torch.nn as nn

class CNN(nn.Module):
    # Specify the components to be created automatically upon
        instantiation
    def __init__(self):
        super(CNN, self).__init__()
        # The first convolutional block
        self.conv1 = nn.Sequential(
            nn.Conv2d(
                in_channels=1, out_channels=16, kernel_size=5,
                    stride=1, padding=2
            ),
            nn.ReLU(),
            nn.MaxPool2d(kernel_size=2),
        )
        # The second convolutional block
        self.conv2 = nn.Sequential(
            nn.Conv2d(16, 32, 5, 1, 2),
            nn.ReLU(),
            nn.MaxPool2d(2),
        )
        # The final fully connected layer which outputs 10
            classes
        self.out = nn.Linear(32 * 7 * 7, 10)

    # Specify the flow of information
    def forward(self, x):
        x = self.conv1(x)
        x = self.conv2(x)
        # Flatten the output to shape (batch_size, 32 * 7 * 7)
        x = x.view(x.size(0), -1)
        output = self.out(x)
        return output
```

Listing 5.12
Defining CNN.

This architecture features the following components:

- Convolutional Blocks:

  - The first block (*conv1*) applies 16 5×5 kernels with a stride of 1 and padding of 2, followed by ReLU activation and max pooling.

  - The second block (*conv2*) increases the depth to 32 kernels.

- Fully Connected Layer:

– The final layer (*self.out*) maps the flattened feature map to 10 outputs, corresponding to the 10-digit classes.

- Data Flow in the *forward* Function:

  – Input data flows sequentially through the two convolutional blocks.

  – The multidimensional feature map is flattened into a vector using the *view()* function before being passed to the fully connected layer.

- Parameter Details:

  – *in_channels* specify the input depth (1 for grayscale images).

  – *out_channels* specify the number of filters, determining the depth of the feature map.

  – *kernel_size* controls the size of the convolutional filter.

Here, the raw outputs of the network (logits) are transformed into probabilities using the softmax function:

$$S(\hat{y}_k) = \frac{e^{\hat{y}_k}}{\sum_{i=1}^{C} e^{\hat{y}_i}},$$

where C is the number of classes. This transformation ensures that the outputs are normalized to sum to 1, facilitating probabilistic interpretation. For example, logits $[1, 2, 3]$ are transformed into probabilities $[0.09, 0.24, 0.67]$. This monotonic transformation preserves the ranking of the logits, ensuring that the class with the highest probability remains the predicted class.

The softmax function can be implemented as follows:

```
import numpy as np
def softmax(x):
    return np.exp(x) / np.sum(np.exp(x), axis=0)

>>> test = [1,2,3]
>>> transformed_test = softmax(test)
>>> print('softmax output:', transformed_test)
softmax output: [0.09003057 0.24472847 0.66524096]
```

Listing 5.13
Implementing Softmax Function.

PyTorch's cost functions, such as CEL, automatically apply the softmax transformation, so it is unnecessary to implement it explicitly during training.

Figure 5.15 shows the process of transforming the original output into probabilities using the softmax function. The transformed outputs are now bounded and sum to one, thus facilitating interpretation. The relative ranking of the outputs also remains the same, so the final prediction is still the third class in this example.

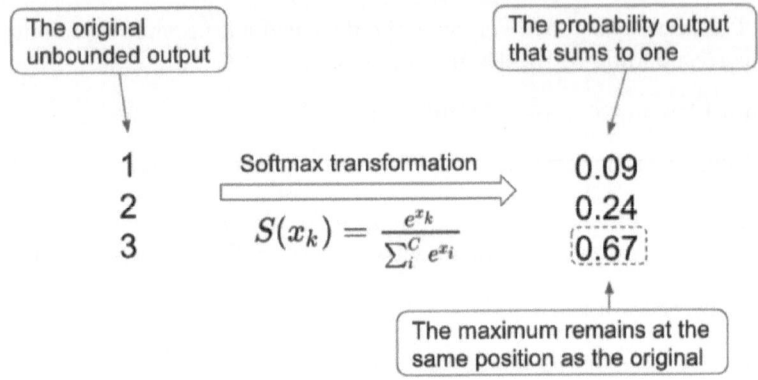

FIGURE 5.15
Transforming the original output into probabilities using the softmax function.

The model can also be instantiated and moved to a GPU for efficient training:

```
device = "cuda" if torch.cuda.is_available() else "cpu"
model = CNN().to(device)
print(model)
# Output
# Using cuda device
# CNN(
#   (conv1): Sequential(
#     (0): Conv2d(1, 16, kernel_size=(5, 5), stride=(1, 1),
      padding=(2, 2))
# (1): ReLU()
# (2): MaxPool2d(kernel_size=2, stride=2, padding=0, dilation=1,
    ceil_mode=False)
#   )
#   (conv2): Sequential(
#     (0): Conv2d(16, 32, kernel_size=(5, 5), stride=(1, 1),
      padding=(2, 2))
#     (1): ReLU()
#     (2): MaxPool2d(kernel_size=2, stride=2, padding=0, dilation
      =1, ceil_mode=False)
#   )
#   (out): Linear(in_features=1568, out_features=10, bias=True)
# )
```

Listing 5.14
Moving to GPU.

When the neural network architecture starts to scale up and become complex, it is often helpful to print out the architecture for better clarification of its composition. In the code snippet below, we resort to the *torchsummary* package to ease the visualization task by passing in the size of an input entry. The output shows the model architecture from top to bottom, with each layer sequentially suffixed by an integer. The output shape and number of

parameters in each layer are also provided, which shows that there are a total of 28,938 (trainable) parameters used in the model. Note that we do not
have any non-trainable parameters; this often relates to the level of model
fine-tuning when performing transfer learning.

```
from torchsummary import summary
summary(model, input_size=(1, 28, 28))
# Output
----------------------------------------------------------------
        Layer (type)            Output Shape          Param #
================================================================
          Conv2d-1           [-1, 16, 28, 28]              416
            ReLU-2           [-1, 16, 28, 28]                0
       MaxPool2d-3           [-1, 16, 14, 14]                0
          Conv2d-4           [-1, 32, 14, 14]           12,832
            ReLU-5           [-1, 32, 14, 14]                0
       MaxPool2d-6            [-1, 32, 7, 7]                 0
          Linear-7                   [-1, 10]           15,690
================================================================
Total params: 28,938
Trainable params: 28,938
Non-trainable params: 0
----------------------------------------------------------------
Input size (MB): 0.00
Forward/backward pass size (MB): 0.32
Params size (MB): 0.11
Estimated Total Size (MB): 0.44
----------------------------------------------------------------
```

Listing 5.15
Model Summary.

5.3.3   Defining the cost function

In machine learning, the cost function measures how well a model's predictions align with the actual target labels. For regression problems, the MSE is
a common choice. However, in classification tasks, such as predicting handwritten digits from the MNIST dataset, CEL is often used. CEL is designed to
quantify the discrepancy between predicted probabilities and one-hot-encoded
target labels, making it particularly effective for multi-class classification.

For the ith observation $\mathbf{x}_i$, let the model output a vector of probabilities
$\mathbf{p}_i = [p_{i,1}, p_{i,2}, \ldots, p_{i,C}]$, where C is the total number of classes. Each $p_{i,k}$
represents the predicted probability of the ith observation belonging to class
k. The CEL for a single observation is defined as:

$$Q_i(\mathbf{w}) = -\sum_{k=1}^{C} y_{i,k} \log(p_{i,k}),$$

where $y_{i,k} = 1$ if the true label for observation i belongs to class k, and $y_{i,k} = 0$
otherwise. This formula evaluates the model's performance across all C classes

by computing a weighted sum of the logarithm of predicted probabilities. Only the term corresponding to the correct class contributes to the loss due to the one-hot encoding nature of the target label.

Intuitively, we would expect the predicted probability for the correct class to be as close to 1 as possible and for the incorrect classes to approach 0. In other words, the loss should increase as the predicted probabilities deviate from the true class label. To clarify, consider the following four scenarios for the ith observation and the kth class:

- Correct Class Prediction (High Confidence): When the target label belongs to the kth class (i.e., $y_{i,k} = 1$) and the predicted probability for the kth class is very high (i.e., $p_{i,k} \approx 1$), the cost should be low. This aligns with the model making a confident and correct prediction.

- Correct Class Prediction (Low Confidence): When the target label belongs to the kth class (i.e., $y_{i,k} = 1$) but the predicted probability for the kth class is very low (i.e., $p_{i,k} \approx 0$), the cost should be high. This penalizes the model for failing to correctly identify the class.

- Incorrect Class Prediction (High Confidence): When the target label does not belong to the kth class (i.e., $y_{i,k} = 0$) but the predicted probability for the kth class is very high (i.e., $p_{i,k} \approx 1$), the cost should also be high. This scenario penalizes the model for assigning high confidence to an incorrect class.

- Incorrect Class Prediction (Low Confidence): When the target label does not belong to the kth class (i.e., $y_{i,k} = 0$) and the predicted probability for the kth class is very low (i.e., $p_{i,k} \approx 0$), the cost should be low. This reflects the model correctly ignoring irrelevant classes.

These scenarios illustrate how CEL operates to reward accurate predictions with high confidence while penalizing incorrect or uncertain predictions. By focusing only on the probabilities corresponding to the true class, the loss function can encode and reflect this intuitive behavior.

The CEL for the entire dataset is computed by summing over all N observations:

$$Q(\mathbf{w}) = -\sum_{i=1}^{N} \sum_{k=1}^{C} y_{i,k} \log(p_{i,k}).$$

This formula aggregates the per-observation loss to provide a single scalar value representing the model's goodness of fit across the dataset. Figure 5.16 summarizes the above discussion on the CEL.

Note that to compute CEL, the target labels must be one-hot encoded. For example, if the true label of an image corresponds to the digit 8, the one-hot encoded vector would be:

$$\mathbf{y}_i = [0, 0, 0, 0, 0, 0, 0, 0, 1, 0, 0].$$

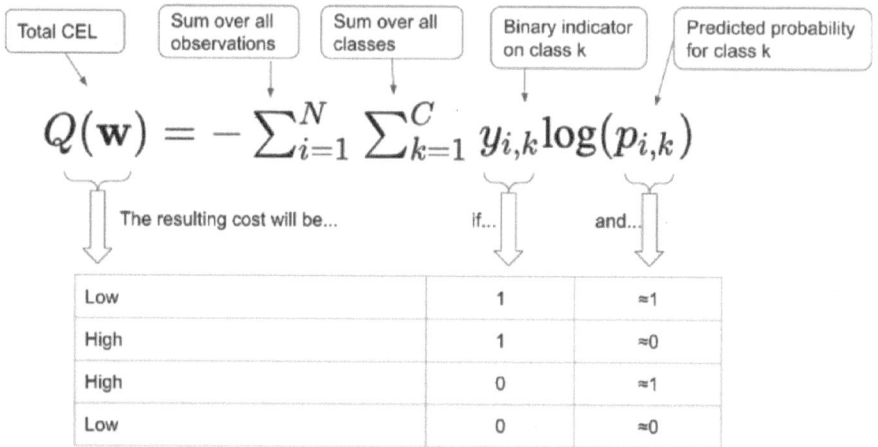

FIGURE 5.16
CEL and its intuition.

Here, the eighth element (corresponding to class 8) is set to 1, while all other elements are set to 0.

The following function implements CEL. It takes the one-hot encoded target vector and the predicted probability vector as inputs.

```
import numpy as np

def cross_entropy(actual, predicted):
    EPS = 1e-15
    predicted = np.clip(predicted, EPS, 1 - EPS)   # Cap
        probabilities to avoid log(0)
    loss = -np.sum(actual * np.log(predicted))
    return loss
```

Listing 5.16
Implementing CEL.

Let us consider a three-class classification problem. The true label is the first class (one-hot encoded as [1, 0, 0]). We evaluate two sets of predictions: one with high confidence for the correct class and one with low confidence.

```
>>> y = np.array([1, 0, 0])
>>> pred_good = np.array([0.8, 0.1, 0.1])
>>> pred_bad = np.array([0.1, 0.6, 0.3])
>>> loss_good = cross_entropy(y, pred_good)
>>> loss_bad = cross_entropy(y, pred_bad)
>>> print('The good prediction loss: {:.4f}'.format(loss_good))
>>> print('The bad prediction loss: {:.4f}'.format(loss_bad))
The good prediction loss: 0.2231
The bad prediction loss: 2.3026
```

Listing 5.17
Calculating CEL.

As expected, the loss is significantly higher for the poor prediction.

PyTorch simplifies the calculation of CEL by automating one-hot encoding and applying the softmax transformation internally. The *torch.nn. CrossEntropyLoss* function can handle these steps, as shown in the following code snippet:

```
import torch
import torch.nn as nn

cost_func = nn.CrossEntropyLoss()  # Initialize the CEL function
y = torch.tensor([0])  # True label as a class index
pred_good = torch.tensor([[2, 0.8, 0.5]])  # Predicted logits
pred_bad = torch.tensor([[1, 6, 0.2]])

loss_good = cost_func(pred_good, y)
loss_bad = cost_func(pred_bad, y)

print('The good prediction loss: {:.4f}'.format(loss_good.item())
    )
print('The bad prediction loss: {:.4f}'.format(loss_bad.item()))
The good prediction loss: 0.4216
The bad prediction loss: 5.0097
```

Listing 5.18
Generating Functional Evaluations.

In this example, the higher loss for the poor prediction aligns with our expectations. CEL thus provides a measure of the model's fit to the training data. Although a lower loss indicates a better fit to the training data, we still need to be cautious about potential overfit. To mitigate this, a regularization term can be added to the cost function, penalizing overly complex models.

5.3.4   Defining the optimization procedure

As mentioned earlier, to update the model parameters, we can employ optimization algorithms, such as SGD and its variants, provided by PyTorch. These optimizers are particularly advantageous because they scale efficiently, enabling the optimization of models with millions of parameters or more. In its basic form, the SGD optimizer requires two key inputs: the model parameters, denoted as $\mathbf{w}$, and the learning rate η, which controls the step size in the parameter space. Once an optimizer is instantiated, we invoke its *step()* method to adjust $\mathbf{w}$ after computing the gradients $\nabla_{\mathbf{w}} J(\mathbf{w})$ using the *backward()* method, where $J(\mathbf{w})$ represents the cost function.

The optimization process is inherently iterative. At each step t, the prediction function $f(\mathbf{x}_i; \mathbf{w}^{(t)})$ generates outputs for a given input $\mathbf{x}_i$ based on the current model parameters $\mathbf{w}^{(t)}$. The cost function $J(\mathbf{w})$ then computes a loss, typically as a function of the predicted values and the ground truth targets y_i. For example, in regression, $J(\mathbf{w})$ could be the MSE:

$$J(\mathbf{w}) = \frac{1}{n} \sum_{i=1}^{n} \left(f(\mathbf{x}_i; \mathbf{w}) - y_i \right)^2 ,$$

where n is the number of training samples. The optimizer updates $\mathbf{w}$ by computing $\nabla_{\mathbf{w}} J(\mathbf{w})$, the gradient of the cost with respect to the weights, and performing a gradient descent step:

$$\mathbf{w}^{(t+1)} = \mathbf{w}^{(t)} - \eta \nabla_{\mathbf{w}} J(\mathbf{w}^{(t)}).$$

This process iterates until convergence, for example, when either the weights $\mathbf{w}$ or the cost $J(\mathbf{w})$ exhibit negligible changes over subsequent iterations. We can also refer to the convergence plot of the loss function to detect if the optimization procedure has reached a stable solution.

The following Python code snippet shows how to set up the optimization procedure using PyTorch's SGD optimizer. Here, the model parameters $\mathbf{w}$ are updated iteratively with a fixed learning rate $\eta = 0.01$:

```
from torch import optim
optimizer = optim.SGD(model.parameters(), lr=0.01)
```

Listing 5.19
Initializing SGD.

5.3.5 Updating model weights

We now consolidate the components introduced earlier into the full training loop. Using the predefined model architecture and a data loader object, the training function implements the SGD optimization algorithm, iterating over the entire training dataset. The training loop updates the model parameters $\mathbf{w}$ batch by batch while tracking the evolution of the cost function. This process is repeated over a fixed number of epochs to iteratively refine the weights and reduce the predictive loss.

```
def train(model, loaders, verbose=True):
    # Sets the model to training mode
    model.train()
    # Extract the total number of images
    total_img = len(loaders['train'].dataset)

    # Iterate through batches
    for batch, (X, y) in enumerate(loaders['train']):
        # Transfer data to the GPU
        X, y = X.to(device), y.to(device)
        # Compute predictions $f(\mathbf{x}^{(i)}; \mathbf{w})$
        pred = model(X)
        # Compute cost $\mathcal{L}(f(\mathbf{x}^{(i)}; \mathbf{w}), y^{(i)})$
        cost = cost_func(pred, y)
```

```
15    # Clear gradients
16    optimizer.zero_grad()
17    # Perform backpropagation to compute $\frac{\partial \
         mathcal{L}}{\partial \mathbf{w}}$
18    cost.backward()
19    # Update $\mathbf{w}$ using gradient descent
20    optimizer.step()
21
22    # Log cost at intervals
23    if verbose:
24        if batch % 100 == 0:
25            loss, current_img_idx = cost.item(), batch * len(
                 X)
26            print(f"cost: {cost:>7f}   [{current_img_idx:>5d
                 }/{total_img:>5d}]")
```

Listing 5.20
Defining the Training Procedure.

In the function, the model prediction $f(\mathbf{x}_i; \mathbf{w})$ is computed for a batch of training samples $\mathbf{X}$. The predictive cost is evaluated using a predefined cost function, which quantifies the discrepancy between the predictions and the ground truth labels $\mathbf{y}$. The *backward()* method calculates the gradients $\frac{\partial \mathcal{L}}{\partial \mathbf{w}}$, which are then used to update the model parameters $\mathbf{w}$ via *optimizer.step()*.

To observe how the cost evolves, we can execute the training function for one epoch:

```
1    >>> train(model, loaders)
2    cost: 2.305853   [    0/60000]
3    cost: 2.070413   [10000/60000]
4    cost: 0.855676   [20000/60000]
5    cost: 0.454673   [30000/60000]
6    cost: 0.475409   [40000/60000]
7    cost: 0.320870   [50000/60000]
```

Listing 5.21
Evolution of Cost.

The decreasing cost values over batches demonstrate that the model is improving its predictions as training progresses.

Besides, we also define a test function to evaluate the model's performance on unseen data. This function computes predictions for test samples and compares them with the ground truth labels to determine the model's accuracy. To ensure efficiency, gradient calculations are disabled during evaluation using the *torch.no_grad()* context:

```
1    def test(model, loaders, verbose=True):
2        # Control the behavior of certain layers by specifying the
             evaluation mode
3        model.eval()
4        # Extract the total number of images to in the test set
5        total_img = len(loaders['test'].dataset)
6        correct = 0
7        # Disable gradient calculation
```

```
 8   with torch.no_grad():
 9       for X, y in loaders['test']:
10           X, y = X.to(device), y.to(device)
11           pred = model(X)
12           correct += (pred.argmax(1) == y).type(torch.float).
                 sum().item()
13       # Add the correct prediction for each batch
14       correct /= total_img
15       if verbose:
16           print(f"Test accuracy: {correct:>0.3f}")
17
18   Using this test function, the performance of the model trained
        with one epoch is evaluated:
19
20   >>> test(model, loaders)
21   Test accuracy: 0.911
```

Listing 5.22
Defining Test Function.

With just one epoch, the model achieves a test accuracy of 91.1%, indicating a reasonable initial fit to the data. To improve performance, we can train the model for multiple epochs, as demonstrated below:

```
 1   >>> num_epochs = 10
 2   >>> for t in range(num_epochs):
 3   >>>     print(f"Epoch {t+1}\n--------------------------------")
 4   >>>     train(model, loaders, verbose=False)
 5   >>>     test(model, loaders)
 6   >>> print("Done!")
 7   Epoch 1
 8   -----------------------------
 9   Test accuracy: 0.911
10   Epoch 2
11   -----------------------------
12   Test accuracy: 0.943
13   ...
14   Epoch 10
15   -----------------------------
16   Test accuracy: 0.980
17   Done!
```

Listing 5.23
Training for More Epochs.

This iterative training procedure results in progressively higher accuracy. After ten epochs, the accuracy reaches 98%, illustrating the benefit of extended training in optimizing the model's performance.

Before we wrap up, let me circle back to the topic of generalization and discuss a few more points about this holy grail.

5.4   More on Generalization

Earlier, we mentioned that SGD is widely used in deep neural networks due
to its strong generalization performance. Interestingly, we did not explicitly
apply any regularization during the training process. Despite this, SGD still
introduces an implicit regularization effect. This effect, combined with the
inherent flexibility of neural network architectures and structured input data
(e.g., image data), helps reduce test error even when training error approaches
zero and the model training continues. This phenomenon can lead to a "double
descent" behavior in the test loss curve, as explored by [15].

To analyze this property more closely, we define a simple neural network
with one fully connected hidden layer. This minimal architecture helps isolate
the impact of model complexity, which we can adjust by varying the number
of nodes in the hidden layer. Below is the definition of the model class. The
nn.Linear() function is used to define the hidden layer and the output layer.
The input size for the hidden layer is set to $1 \times 28 \times 28$, and the output size
corresponds to the ten classes in the MNIST dataset. The size of the hidden
layer, *num_nodes*, is a hyperparameter and is used to allow experimentation
with varying model complexities. The *forward()* function then flattens the
input data before passing it through the fully connected layer.

```
import torch.nn as nn
import torch.nn.functional as F
class SimpleNN(nn.Module):
    # Model initialization codes
    def __init__(self, num_nodes, num_classes=10):
        super(SimpleNN, self).__init__()
        # The single fully connected layer
        self.fc = nn.Linear(1 * 28 * 28, num_nodes)
        # The final fully connected layer which outputs 10
            classes
        self.out = nn.Linear(num_nodes, num_classes)
    # Mode architecture codes
    def forward(self, x):
        # Flatten all dimensions except for the batch size
        x = x.view(x.shape[0], -1)
        x = F.relu(self.fc(x))
        output = self.out(x)
        return output
```

Listing 5.24
Defining a Simple Neural Network.

Next, we need a function to tell us how well the model performs on the
training and test sets. These also give us an indication of the model's generalization performance when we vary the hidden layer's size. In the following
code listing, we define a function called *check_accuracy()* to calculate the predictive accuracy for the entire training or test set, depending on the value of
the input argument. The codes largely follow the same structure as the *test()*

function in the previous example, except that we use the input image loader based on a user-specified parameter. This saves us from writing two similar functions for both training and test sets.

```python
def check_accuracy(model, loaders, verbose=True, type="train"):
    # Set the model to evaluation mode
    model.eval()

    # Extract the total number of images
    total_img = len(loaders[type].dataset)
    correct = 0

    # Disable gradient calculation as backpropagation is not
        needed
    with torch.no_grad():
        for X, y in loaders[type]:
            X, y = X.to(device), y.to(device)

            # Compute model predictions
            pred = model(X)

            # Add the correct prediction for each batch
            correct += (pred.argmax(1) == y).type(torch.float).
                sum().item()

    # Calculate accuracy as the proportion of correct predictions
    correct /= total_img
    if verbose:
        print(f"{type} accuracy: {correct:>0.3f}")
    return correct
```

Listing 5.25
Checking Model Accuracy.

We are now ready to assess the impact of model complexity on the predictive accuracy for both the training set and the test set. In the following code listing, we start with two nodes in the hidden layer and gradually double the size until 4096. The reason for such an extensive range is to see how the model performs as the model complexity increases even after passing the interpolating threshold. We also set a total training budget of 500 epochs in *num_epochs* to ensure that the training procedure can reach convergence, that is, the model parameters do not change much. Upon exhausting the training budget, we record the final prediction accuracy for the training set in *final_train_acc* and the test set in *final_test_acc*. Since running the full codes takes quite some time, we store these two variables in text files via the pickle package so that we do not need to start from scratch if the running gets disrupted.

```python
import pickle
from torch import optim

# Set the device to GPU if available, otherwise use CPU
device = "cuda" if torch.cuda.is_available() else "cpu"

# Set the total number of epochs for training
```

```
 8  num_epochs = 500
 9
10  # Initialize lists to store final training and test accuracies
11  final_train_acc = []
12  final_test_acc = []
13
14  # Define the range of nodes in the hidden layer for
         experimentation
15  num_nodes = [2, 4, 8, 16, 32, 64, 128, 256, 512, 1024, 2048,
         4096]
16
17  for n_node in num_nodes:
18      # Print the current number of nodes being evaluated
19      print(f"------NODE: {n_node}-----")
20
21      # Initialize the model with the specified number of nodes
22      model = SimpleNN(num_nodes=n_node)
23
24      # Define the cost function
25      cost_func = nn.CrossEntropyLoss()
26
27      # Define the optimizer using SGD
28      optimizer = optim.SGD(model.parameters(), lr=0.01)
29
30      # Start the training process
31      for t in range(num_epochs):
32          # Print the epoch number every 10 epochs
33          if t % 10 == 0:
34              print(f"Epoch {t}")
35
36          # Train the model on the current dataset
37          train(model, loaders, verbose=False)
38
39      # Save the final training accuracy
40      final_train_acc.append(check_accuracy(model, loaders, type="
             train"))
41
42      # Save the final test accuracy
43      final_test_acc.append(check_accuracy(model, loaders, type="
             test"))
44
45      # Save the training accuracies to a text file
46      with open(f"final_train_acc_{n_node}.txt", "wb") as fp:
47          pickle.dump(final_train_acc, fp)
48
49      # Save the test accuracies to a text file
50      with open(f"final_test_acc_{n_node}.txt", "wb") as fp:
51          pickle.dump(final_test_acc, fp)
```

Listing 5.26
Running Multiple Experiments.

In order to align our focus on the loss instead of accuracy and observe where the interpolating threshold occurs, we will use one to minus off the accuracy to derive the misclassification rate and round it off to three decimals. This is achieved via list comprehension after loading the saved training and test accuracies for different sizes of the hidden layer (see the code listing below). This allows us to observe the changes in training and test misclassification rates in the y-axis as the model complexity increases in the x-axis. In addition to plotting the two misclassification rate curves, we also draw the interpolating threshold by detecting the first occurrence of zero value in the train variable via the *index* method.

```python
import matplotlib.pyplot as plt
import numpy as np

# Load the saved training accuracies from the file
with open("final_train_acc_4096.txt", "rb") as fp:
    train = pickle.load(fp)

# Load the saved test accuracies from the file
with open("final_test_acc_4096.txt", "rb") as fp:
    test = pickle.load(fp)

# Convert accuracy to misclassification rate and perform rounding
train = [np.around(1 - i, 3) for i in train]
test = [np.around(1 - i, 3) for i in test]

# Define the x-axis values corresponding to the number of nodes
x_axis = range(len(num_nodes))

# Plot the misclassification rate for the training set
plt.plot(x_axis, train, label="train")

# Plot the misclassification rate for the test set
plt.plot(x_axis, test, label="test", ls="--")

# Add labels for the x-axis ticks
plt.xticks(x_axis, num_nodes)

# Detect the first occurrence of zero misclassification and add a
#     vertical line
plt.axvline(x=x_axis[train.index(0)], ls="--")

# Annotate the interpolation threshold
plt.text(
    x_axis[train.index(0)] - 0.2,
    0.1,
    "Interpolation threshold",
    fontsize=10,
    color="gray",
    rotation=90,
    rotation_mode="anchor",
)

# Add labels to the axes
plt.xlabel("# hidden nodes")
plt.ylabel("Misclassification rate")

# Add a legend to the plot
plt.legend()
```

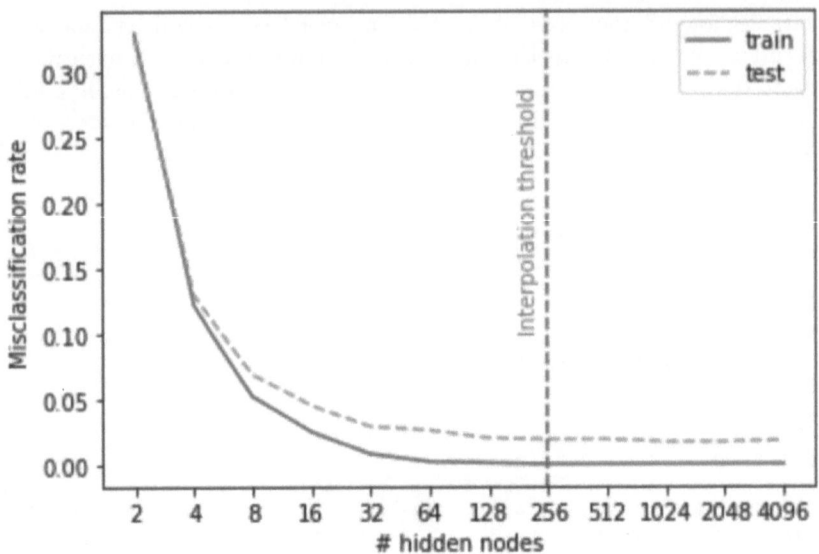

FIGURE 5.17
The misclassification curves for both training and test sets as a function of the number of nodes in the hidden layer. Both curves exhibit a decreasing trend even if the model continues to become more complex, which corresponds to the modern generalization theory. In particular, when the model starts to perfectly interpolate the training data with 256 hidden nodes, the misclassification rate for the test set keeps decreasing with a larger hidden layer in size.

```
49   # Display the plot
50   plt.show()
```

Listing 5.27
Analyzing Model Performance.

Running the codes above will produce Figure 5.17. Recall that each point represents the final training or test misclassification rate upon completing the 500 epochs of the training procedure. The misclassification rate curves could thus be considered a reflection of the model's performance upon convergence. Both training and test curves exhibit a uniform decreasing trend, with the model reaching 100% training accuracy and perfectly interpolating the training data when there are 256 nodes in the hidden layer. Interestingly, the test misclassification rate decreases even beyond the interpolating threshold, which is counterintuitive based on the classical generalization theory and an excellent example of the modern generalization theory.

Besides, the "double-descent" curve is immediately obvious, and this could be due to the increasing gap between different model complexities so that the second descent is obscured. Another possibility is that the implicit regularization effect of the SGD optimization algorithm is strong enough to alleviate the "double-descent" phenomenon. Proper regularization can hide the "double-descent" phenomenon and accelerate learning toward a generalizing model.

5.4.1 Multiple global minima

We have observed so far that the solutions (i.e., the model weights upon convergence) returned by training the model using SGD exhibit good generalization performance in the test set, and in general, there are multiple solutions with such equally good performance. In other words, there are multiple equally good local minima in the loss surface, and using SGD tends to land in these solutions that display good generalization behavior. However, when we are not using SGD, the model may converge to a bad solution for the same model configuration, largely due to the lack of the implicit regularization effect.

Let us analyze this specific attribute in detail. Figure 5.18 shows two different types of local (and potentially global) minima based on the training loss curve: the sharp valley and the flat basin. The sharp valley is where the best solution lies. It is the global minimum, although the figure shows that this solution is located in a narrow valley and is thus sensitive. When training neural networks to reach this region, the training procedure needs to be delicate enough to reach the bottom of the valley. In comparison, the multiple local minima living in the flat basin area on the right are much less sensitive to the location of the weights within that area, where the error changes slowly.

These two distinct error profiles are commonly encountered in many deep learning problems. The overall recommendation is to train models that reach the flat area with many equally good local minima or basins of attraction. These models tend to be more regularized and generalize better to the unseen test set (see Figure 5.19). Despite a similar error in the training set, as represented by a zero misclassification rate, the corresponding test error may vary depending on the specific solution zone. Specifically, a wide and flat area has a lower test error than the one obtained in the sharp zone. One particular approach brought up recently is stochastic weight averaging, proposed by [15], which averages a few "sharp" solution weights to arrive at a final "flat" solution within the basin zone. This technique belongs to ensemble learning, where multiple models are averaged (in terms of model weights) to generate the final prediction. We will cover a specific ensembling technique called bootstrap in a later chapter in the book series on regularization via data.

5.4.2 Best versus worst minimum

We have learned so far that, regarding the solution space (model weights upon convergence), there are multiple equally good local minima (in terms of training error) when the model is trained to interpolate the training data given a fixed model complexity and across different levels of model complexity.

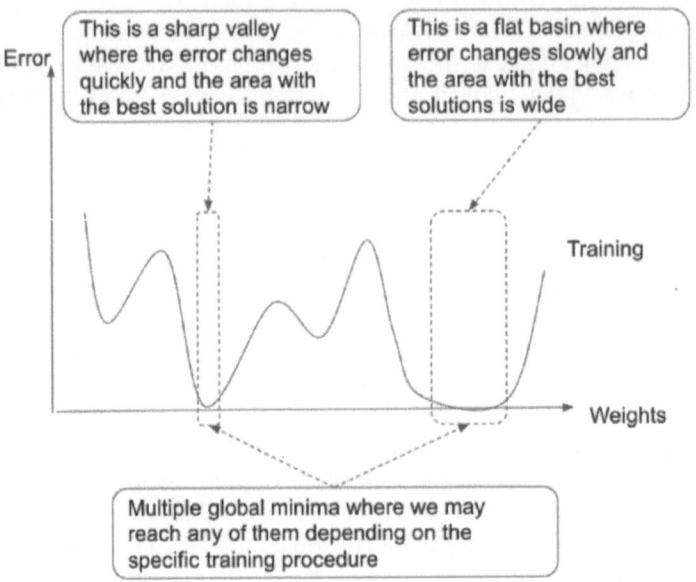

FIGURE 5.18
Illustrating the two types of local or potentially global minima. On the left shows a sharp valley where the error curve changes drastically upon reaching the bottom of the valley. On the right shows a set of equally good global minima along a flat basin where the error changes slowly.

Out of these many equally good solutions, some tend to generalize well to the test set, for example, via implicit regularization when using SGD. Some tend to generalize poorly due to overfitting. Suppose we were to characterize the solution space of local minima for a fixed level of model complexity. In that case, we can obtain an upper bound (i.e., worst global minimum) and a lower bound (i.e., best global minimum). The upper bound would be the subject of study under the classical generalization theory that highlights a high test error due to overfitting. On the other hand, the lower bound reflects the best model that can be achieved via implicit or explicit regularization.

Let us look at this perspective more closely. Figure 5.20 illustrates the upper and lower bounds of the test error as a function of the model complexity, plus the training error as a reference. The training error decreases as the model becomes complex under both classical and modern generalization views. However, the behaviors differ in the test error. The test error obtained under the classical generalization theory forms the upper bound of all possible test errors for a given level of model complexity due to overfitting. The increasing test error shows the importance of regularizing the (effective) model complexity to control for the excessive sensitivity during model fitting. The lower bound is the empirically observed best global minimum across different

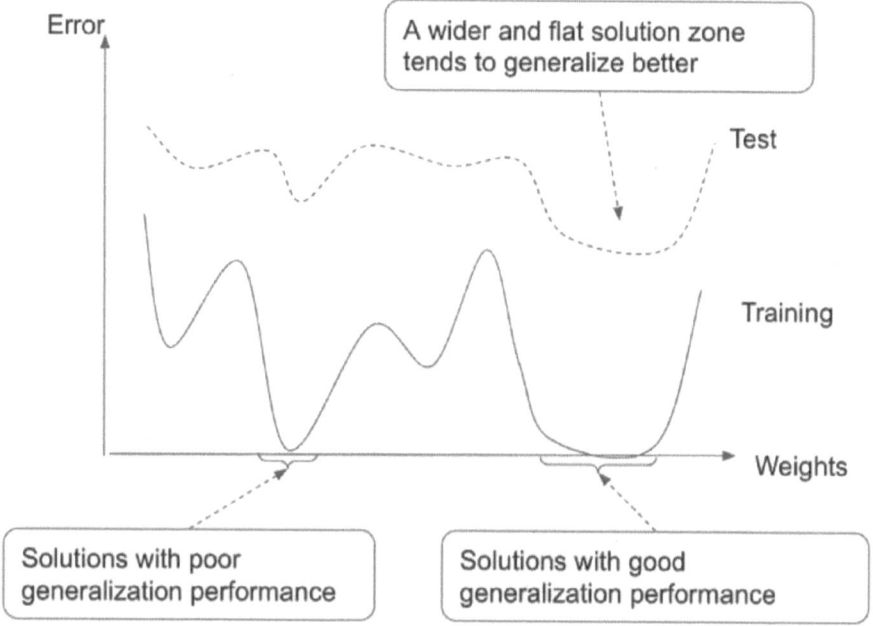

FIGURE 5.19

Good and bad solution zones in the training and test error curves. There are multiple solutions where the weights can interpolate the training data and produce zero misclassification rate. However, solutions living within the wide and flat region tend to better generalize to the test set as represented by a lower misclassification rate.

levels of model complexity. We can obtain the lower bound using implicit regularization (i.e., SGD), explicit regularization (i.e., penalizing the norm of the weights), or both.

Besides, we also notice that the gap between the upper and lower bounds of the possible test errors increases as the model becomes complex. This shows that the estimation error tends to get large as the model class becomes more complex, which leads to a bigger hypothesis space.

When the model is less complex and small in size, we will obtain a unique or very similar local minimum in the test error as the model is trained to minimize the training error. However, the gap increases between the best and worst minimum as the model becomes flexible and complex. It is thus essential to apply proper regularization for large and complex models such as deep neural networks to obtain the best possible minimum and generalize well to the future unseen test set.

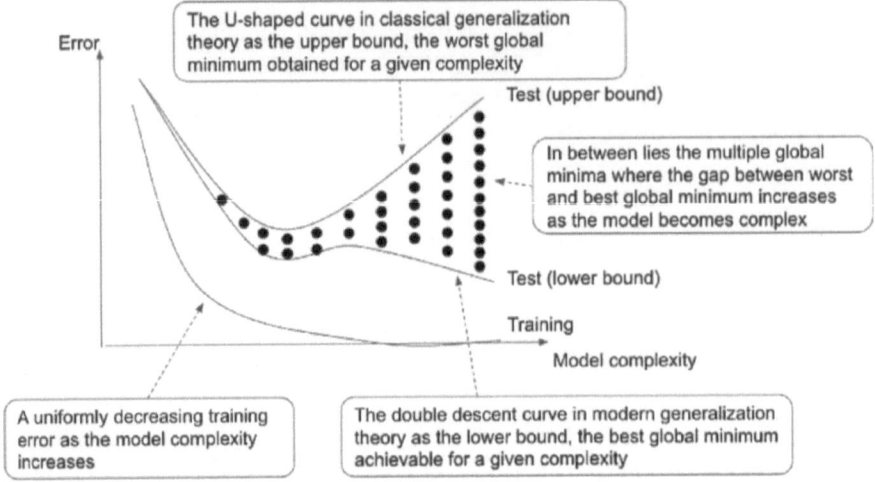

FIGURE 5.20

The worst and best minimum across different levels of model complexity. The upper bound is the worst local or global minima obtained under classical generalization theory when no regularization mechanism is used. The lower bound serves as the best obtainable local or global minima under the modern generalization theory. The gap between the upper bound and the lower bound increases as the model becomes complex, showing an increase in the estimation error.

5.5 Summary

This chapter provides an in-depth introduction to deep neural networks, starting with the fundamental principles behind function approximation. It explains how adding features and controlling architectural complexity enable a model to better approximate an underlying function, echoing the intuition from polynomial approximations. We mentioned the universal approximation theorem, which states that, given sufficient width and properly chosen weights, neural networks can approximate any continuous function on a compact domain. This sets the stage for a deeper dive into how networks, starting from simple perceptrons, evolve into complex architectures capable of modeling highly nonlinear relationships.

The discussion then shifts to MLPs, emphasizing the role of activation functions like the ReLU in introducing nonlinearity. Detailed examples illustrate how stacking perceptrons transform a linear model into one that can approximate complex functions through the composition of linear transformations and nonlinear activations. A comparison between shallow and deep networks highlights that while shallow networks increase expressivity linearly with additional nodes, deep networks can achieve exponential growth in expressivity through additional layers. This difference is exemplified by the "sawtooth" network model, which demonstrates how increasing depth can yield a dramatic increase in the number of linear segments that a network can represent.

We further explore the mechanics of training these networks by introducing automatic differentiation and gradient-based optimization, covering how the chain rule is employed during backpropagation to compute gradients across multiple layers. We also compare forward and backward modes of matrix multiplication in terms of computational cost and memory efficiency. These concepts are crucial for understanding how modern deep learning frameworks, such as PyTorch, enable efficient training of complex models through SGD and related optimization algorithms.

We also describe the process of training a CNN on the MNIST dataset. This section walks through the stages of data preparation, model definition, cost function formulation (using CEL), and optimization. Code examples illustrate the complete training loop, from downloading the dataset and visualizing samples to defining the CNN architecture and evaluating model performance. The section demonstrates how high accuracy can be achieved through iterative training.

Finally, we add more insight into generalization in deep neural networks. We show that, despite perfectly interpolating the training data, the model continues to improve its performance on unseen data when increasing its complexity. By comparing sharp and flat minima, we highlight that flat regions in the loss landscape tend to generalize better, underscoring the importance of both model architecture design and training strategy in achieving robust performance.

6

A Concluding Perspective

Deep learning has emerged as a transformative suite of tools in the field of machine learning and artificial intelligence, enabling significant breakthroughs in a variety of complex predictive tasks. From achieving superhuman performance in image recognition to mastering games like Go and enabling sophisticated natural language understanding, deep neural networks have demonstrated strong approximation capability to learn intricate patterns from large datasets. This success is largely attributed to their capacity to model high-dimensional and nonlinear relationships, where multiple layers of abstraction can be applied to facilitate rich and effective learning.

Despite these practical achievements, the theoretical understanding of why deep learning generalizes so well remains a subject of intense research and debate. As discussed earlier on the modern perspective of generalization, one of the most perplexing and fundamental questions is how and why deep neural networks, whose number of parameters often vastly exceeds the number of training samples, generalize effectively to new, unseen data in many cases. This is counterintuitive from the classical perspective of statistical learning theory, which suggests that models with such high capacity tend to overfit the training data and learn the noise rather than the signal, thus performing poorly on future unseen samples.

The generalization paradox observed in many deep learning tasks challenges the traditional notions of bias-variance trade-off and model capacity. It seems that classical theories, which can be based on concepts like the Vapnik-Chervonenkis (VC) dimension and Rademacher complexity, fail to provide satisfactory explanations for the empirical success of overparameterized neural networks. These complex models can often perfectly fit training data (also called perfect interpolation), even when labels are randomized, yet exhibit strong generalization performance on real-world tasks.

In this chapter, we aim to summarize the theoretical underpinnings of generalization in deep learning that we have covered so far. We will explore the historical foundations that laid the groundwork for understanding neural networks, including their expressivity and trainability. By examining key theoretical results, such as the Universal Approximation Theorem, and analyzing the optimization landscapes of neural networks, we will build a foundation for discussing the mechanisms that may explain generalization in modern deep learning. We will also highlight the main theoretical contributions and outline the open challenges that continue to motivate research in this area.

DOI: 10.1201/9781003511601-6

6.1   Early Theoretical Foundations

The theoretical study of neural networks has its roots in the quest to understand their ability to approximate complex functions and the dynamics of their training processes. Two primary pillars underpin this foundational work: expressivity, which describes the capacity of neural networks to represent a wide class of functions, and trainability, which denotes the ability to effectively optimize neural networks and obtain good parameter estimates, despite their non-convex loss landscapes. Essentially, the former refers to the approximation error of the specified hypothesis class, and the latter refers to the estimation error.

6.1.1   Expressivity of neural networks

The expressivity, or representational capacity, of neural networks refers to their ability to approximate complex functions drawn from a given function class. Understanding expressivity is crucial because it sets the stage for what neural networks can, in principle, learn given sufficient data and appropriate training algorithms. As we have shown in Chapter 5, the practical realization of this capacity involves intricate trade-offs between the depth and width of a neural network. For example, deeper networks can represent functions with exponentially more linear regions than shallow ones, thus leading to more efficient approximations of highly nonlinear mappings.

Let us review the universal approximation theorem in more detail.

6.1.1.1   The universal approximation theorem

One of the cornerstone results in neural network theory is the Universal Approximation Theorem, which asserts that a feedforward neural network with a single hidden layer can approximate any continuous function on a compact subset of $\mathbb{R}^n$ to any desired degree of accuracy, provided that the network has enough hidden units and uses a suitable activation function. This result is formalized in the following theorem.

Theorem (Cybenko, 1989; Hornik, Stinchcombe, and White, 1989): Let $\sigma : \mathbb{R} \to \mathbb{R}$ be a continuous, non-constant activation function. For any continuous function $f : K \to \mathbb{R}$ defined on a compact set $K \subset \mathbb{R}^n$, and for any $\varepsilon > 0$, there exists an integer M and real constants $a_i \in \mathbb{R}^n$, $b_i \in \mathbb{R}$, and $c_i \in \mathbb{R}$ for $i = 1, 2, \ldots, M$, such that the neural network function:

$$N_M(x) = \sum_{i=1}^{M} c_i \, \sigma(a_i^\top x + b_i)$$

satisfies:

$$\sup_{x \in K} |f(x) - N_M(x)| < \varepsilon.$$

The proof of this theorem relies on the density of neural network function spaces in the space of continuous functions on compact sets. A key element of the argument is the Stone-Weierstrass Theorem, which says that any algebra of functions that contains the constants and separates points is dense in the space of continuous functions $C(K)$. Here, the algebra consists of functions of the form $\sigma(a^\top x + b)$, which are closed under addition and scalar multiplication. These functions separate points in the domain K because, for any two distinct points $x, y \in K$, there exists a choice of a and b such that $\sigma(a^\top x + b) \neq \sigma(a^\top y + b)$. And, by the Stone-Weierstrass Theorem, a finite linear combination of these functions can approximate any continuous function on K, up to an arbitrary degree of accuracy.

The theorem holds for a variety of activation functions, including the sigmoid function $\sigma(z) = \frac{1}{1+e^{-z}}$ and the hyperbolic tangent function $\sigma(z) = \tanh(z)$. In particular, the Rectified Linear Unit (ReLU) activation function, $\sigma(z) = \max(0, z)$, which is widely used in modern neural networks, can also satisfy the conditions of the theorem with appropriate modifications. These results demonstrate the versatility and power of neural networks as universal approximators.

While the Universal Approximation Theorem is profound in its implications, it is important to recognize its limitations. First, the theorem is non-constructive, meaning it guarantees the existence of parameters a_i, b_i, and c_i but does not provide a method for finding these optimal parameters. Moreover, it does not specify how many nodes M in the hidden layer are needed to achieve a given approximation error ε. Additionally, the theorem applies only to continuous functions defined on compact sets, and its extension to unbounded domains requires additional considerations.

As an example, consider the example of approximating a continuous function $f : [0, 1] \to \mathbb{R}$. As covered in Chapter 3, polynomials are flexible functions that can be used to approximate an arbitrarily shaped function. This observation has also been formalized in the Weierstrass Approximation Theorem, in that f can be uniformly approximated by polynomial functions. Since neural networks can use polynomial functions as activation functions to represent polynomials, one might conjecture that such networks can approximate f. However, polynomial activation functions do not satisfy the requirements of the Universal Approximation Theorem. If the activation function is a polynomial, then any neural network built will itself be a polynomial function, due to the compositions of affine transformations and polynomial functions happening in the middle layers.

In contrast, when using non-polynomial activation functions (such as sigmoid, tanh, or ReLU), the network is capable of generating a much richer and more flexible class of functions, allowing it to approximate any continuous function arbitrarily well. This non-polynomial condition is essential to ensure that the network's representational capacity is sufficient to capture the complexity of arbitrary continuous functions, which is why polynomial activations are excluded from the standard formulation of the Universal Approximation Theorem.

For example, when using a sigmoidal activation function such as $\sigma(z) = \frac{1}{1+e^{-z}}$, we can construct a neural network:

$$N_M(x) = \sum_{i=1}^{M} c_i\, \sigma(k_i x + b_i),$$

where k_i and b_i are parameters that control the shape and position of each sigmoid. When these parameters are carefully trained and optimized, the network can approximate f to any desired precision.

The theorem also motivates constructive approximation techniques for specific classes of functions. For example, we know that Fourier series can approximate periodic functions that display regular patterns such as seasonality, and neural networks can approximate these series by tuning the parameters to mimic sinusoidal components. Similarly, wavelet decompositions offer a basis for localized function approximation, which neural networks can learn to capture localized features of complex functions.

Extensions of the Universal Approximation Theorem to deep networks also reveal additional insights. For example, deep neural networks with ReLU activations are also universal approximators, but their efficiency in representing certain functions increases with depth. Indeed, as we have shown earlier, functions that require exponentially many neurons in a shallow network can be represented with polynomially many neurons in a deep network. These results demonstrate the expressive power of depth in modern neural network architectures.

In summary, the Universal Approximation Theorem establishes the theoretical foundation for the representational capacity of neural networks. It ensures that, given sufficient capacity and appropriate activation functions, neural networks can approximate any continuous function on a compact domain. However, practical considerations such as optimization, generalization, and computational efficiency remain critical areas of research in terms of both theory and practice.

6.1.1.2   Tradeoff between depth and width

While the Universal Approximation Theorem establishes the capability of neural networks to approximate any continuous function, it does not address additional questions such as the efficiency or rate of approximation, nor does it consider how the architecture's depth and width affect this efficiency. Subsequent research has focused on quantifying the approximation error in terms of the network size (number of neurons) and depth (number of layers), aiming to better understand the trade-offs between these architectural parameters, often called hyperparameters.

Barron (1993) advanced the theoretical understanding by providing quantitative bounds on the approximation error of feedforward neural networks with sigmoidal activation functions. He demonstrated that, for certain classes of functions, the approximation error decreases at a rate of $O(1/\sqrt{M})$, where M is the number of hidden units in the network. This result offers a more precise characterization of how the network size affects its approximation capabilities. This result is formalized as follows.

Theorem (Barron, 1993): Let f be a function defined on $[0,1]^n$ such that its Fourier transform $\hat{f}(\omega)$ satisfies:

$$\int_{\mathbb{R}^n} \|\omega\|_1 \, |\hat{f}(\omega)| \, d\omega < \infty.$$

Then, for any $M \in \mathbb{N}$, there exists a neural network $N_M(x)$ with one hidden layer and M hidden units using sigmoidal activation functions such that:

$$\|f - N_M\|_{L^2([0,1]^n)} \le \frac{C}{\sqrt{M}},$$

where C is a constant depending only on f but not on M.

Barron's theorem establishes that the approximation error in the L^2 norm decreases inversely with the square root of the number of hidden units, provided that f has finite first moment of its Fourier transform weighted by $\|\omega\|_1$. The condition on $\hat{f}(\omega)$ implies that f belongs to a class of functions with bounded variation and certain smoothness properties, which means it should not be too wiggly in shape. This result provides a quantitative rate at which the approximation error decreases with the number of hidden units. The $O(1/\sqrt{M})$ rate indicates that increasing the number of neurons leads to a predictable improvement in the approximation accuracy. This contrasts with the Universal Approximation Theorem and provides a quantitative measure on the approximation rate as the neural network becomes more complex (more hidden nodes).

Barron's bounds also suggest that modern neural networks can mitigate the curse of dimensionality for certain function classes by proper architectural design. Although traditional approximation methods may require exponentially many parameters in high dimensions, neural networks can achieve acceptable approximation errors with a polynomial number of neurons when designing the neural network to be deep rather than wide, since the depth (number of layers) of a neural network plays a crucial role in its expressive power. Indeed, recent studies have demonstrated that deep networks can represent certain functions exponentially more efficiently than shallow networks, highlighting the importance of depth in neural network architectures.

We introduce one such representative work proposed by Telgarsky (2016) in the recent literature.

Theorem (Telgarsky, 2016): For any $L \in \mathbb{N}$, there exists a function $f_L : [0,1] \to [0,1]$ that can be computed by a ReLU network of depth L with a constant width (independent of L), such that:

- The function f_L oscillates 2^L times over the interval $[0,1]$.

- Any ReLU network of depth less than L that approximates f_L within a constant error $\varepsilon < 1/2$ requires a width at least proportional to 2^L.

This theorem demonstrates that certain functions can be efficiently represented by deep networks but not by shallow ones unless the shallow networks have exponentially many neurons.

Note that the constructed function f_L is designed to perform hierarchical computation, where each additional layer in the network captures increasingly finer details. Specifically, f_L can be thought of as a sawtooth function (covered in Chapter 5) with an exponential number of oscillations, which naturally aligns with the compositional structure of deep networks. This result suggests that depth contributes fundamentally to a network's expressive power. We can also say that deep networks can represent functions that are empirically intractable for shallow networks of practical size, which further indicates that certain tasks inherently benefit from deeper architectures.

In addition, the results emphasize a trade-off between depth and width. Although increasing the number of layers can exponentially reduce the required number of neurons for specific functions, deeper networks may be more challenging to train due to issues like vanishing or exploding gradients, echoing the concept of increasing estimation error for a larger hypothesis class, as covered in Chapter 2. In contrast, shallow networks may require impractically large widths to achieve the same approximation power.

These findings also inform network design in practical applications. For tasks where the target function exhibits a hierarchical or compositional structure, deep networks are more suitable. Examples include image recognition and natural language processing, where features are naturally organized in layers of abstraction.

The exploration of approximation rates and depth-width trade-offs in neural networks also connects to classical approximation theory. In polynomial approximation, for instance, the degree of the polynomial plays a role similar to the depth of a neural network. In particular, higher-degree polynomials can approximate more complex functions, and deeper networks can capture more intricate patterns. These studies extend the understanding of the expressive capabilities of neural networks beyond the mere universal approximation. By quantifying the architectural considerations (number of neurons and layers) required to achieve a certain level of approximation accuracy, they offer a more nuanced perspective on the learning capacity of a neural network.

6.1.2 Trainability and optimization landscapes

Beyond the functional expressivity that determines the approximation error, the practical success of neural networks also depends critically on the ability to train them effectively, which concerns the estimation error. For deep neural networks, this involves navigating the high-dimensional and often non-convex optimization landscapes defined by their loss functions. The training process seeks to find parameter values that minimize a predefined loss function $L(\theta)$, where θ represents all the network parameters, including weights and biases across all layers.

Note that the loss functions associated with neural networks are typically non-convex with respect to the parameters θ. As discussed earlier, such non-convexity introduces several optimization challenges:

- Firstly, the presence of multiple local minima can complicate the optimization process. In a non-convex landscape, there may be multiple minima that are locally optimal but not globally optimal. Gradient-based optimization algorithms, such as SGD, are likely to converge to these local minima, potentially resulting in suboptimal performance.

- Secondly, saddle points pose a significant obstacle. A saddle point is a critical point where the gradient vanishes, but the associated Hessian matrix has both positive and negative eigenvalues, indicating directions of both ascent and descent. In high-dimensional parameter spaces, saddle points are more prevalent than local minima (see more details in Dauphin et al. [2014]). Gradient descent algorithms can thus be easily trapped in flat regions near saddle points, leading to slow convergence or stagnation during the optimization process.

- Thirdly, the optimization landscape may contain plateaus and flat regions where the gradient is small in magnitude over nearby regions of the parameter space. These areas can cause gradient-based methods to make minimal progress, thus prolonging the training process and making it sensitive to hyperparameter choices like the learning rate.

Despite these theoretical challenges, neural networks are trained successfully in practice using variants of SGD. This raises the question: why do these algorithms work effectively despite the non-convexity of the loss landscape?

6.1.2.1 Insights into loss surface geometry

Researchers have investigated the geometric properties of the loss surfaces of neural networks to understand the empirical success of gradient-based optimization. Two significant contributions in this area are the loss surface analysis by Choromanska et al. and the study of properties of deep linear networks by Kawaguchi.

Choromanska et al. drew analogies between the loss surfaces of deep neural networks and the energy landscapes of spin glasses, which are disordered magnetic systems studied in statistical physics. The key idea is that the complexity of the loss landscape can be analyzed using probabilistic methods inspired by spin glass theory. In spin glasses, the energy landscape is characterized by a multitude of local minima separated by energy barriers. Similarly, the loss surface of a neural network can be thought of as a high-dimensional surface with numerous local minima corresponding to different configurations of the network parameters.

To facilitate analysis, Choromanska et al. considered a simplified model of neural networks where the weights are assumed to be random variables with certain statistical properties. By applying results from the spin glass theory, they developed the following insights into the distribution of local minima:

- High-Dimensional Concentration: In high-dimensional parameter spaces, random projection theory and the concentration of measure phenomenon imply that the values of the loss function at different local minima are highly concentrated around their mean. This means that most local minima have loss values close to the global minimum, and reaching any of these equally good local minima would satisfy the termination condition of an optimization procedure.

- Value of Local Minima: The expected loss at a local minimum is only slightly worse than the loss at the global minimum. Mathematically, if $L_{\min}$ denotes the global minimum loss and L_{local} denotes the loss at a local minimum, then:

$$\mathbb{E}[L_{\text{local}}] \approx L_{\min} + \delta$$

where δ is a small value that decreases with the size of the network. This means that the resulting model, which identifies any of the equally good local minima, will not be too far away from the global minimum in terms of the loss value.

- Connectivity of Minima: There exist paths in the parameter space that connect different local minima without traversing regions of high loss. This suggests that the loss surface may have a connected structure, allowing optimization algorithms to move between minima efficiently. This is the path that might possibly take us to another local minimum without a bumpy ride.

These findings imply that, in large-scale neural networks, the majority of local minima are nearly as good as the global minimum in terms of training loss. Consequently, gradient-based methods are likely to find acceptable solutions that generalize well, even if they do not reach the exact location of the global minimum.

6.1.2.2 Properties of deep linear networks

Although real-world neural networks are nonlinear due to the use of activation functions such as ReLU or sigmoid, studying deep linear networks still provides valuable theoretical insights. A deep linear network is a neural network where all activation functions are linear, identity mappings. The function computed by such a network with L layers can be expressed as:

$$f(x; \theta) = W^{(L)} W^{(L-1)} \cdots W^{(1)} x,$$

where $W^{(l)}$ denotes the weight matrix of layer l.

Kawaguchi analyzed the optimization landscape of deep linear networks under the squared loss function:

$$L(\theta) = \frac{1}{2} \sum_{i=1}^{N} \|y_i - f(x_i; \theta)\|^2,$$

where $\{(x_i, y_i)\}_{i=1}^{N}$ are the training samples.

The key results from their analysis are:

- **All Local Minima are Global Minima:** Every local minimum of the loss function $L(\theta)$ corresponds to a global minimum. That is, if θ^* is a local minimum, then $L(\theta^*) = L_{\min}$.

- **All Non-Global Critical Points are Saddle Points:** Any critical point θ that is not a global minimum is a saddle point. This means that in any non-minimal critical point, there exists at least one direction in parameter space along which the loss function can decrease.

These findings suggest that in deep linear networks, optimization algorithms are not hindered by poor local minima, as all local minima are globally optimal, and identifying any of these local minima will suffice. This simplifies the optimization landscape compared to general non-convex functions. In addition, the presence of saddle points explains why optimization can still be challenging. Algorithms can become trapped in saddle regions where gradients are small, leading to slow convergence.

While real neural networks are nonlinear, the insights from linear networks suggest that the optimization landscapes of deep networks may be more friendly than traditionally feared. Empirical studies have observed similar phenomena in nonlinear networks, supporting this perspective.

6.1.2.3   Why optimization works in practice

Combining the insights from [4] and [8], we can better understand why gradient-based optimization methods like SGD are effective for training deep neural networks. We can summarize these findings into the following four perspectives:

- **High-Dimensional Geometry Aids Optimization:** In high-dimensional parameter spaces, the concentration of measure and the abundance of near-global minima increase the likelihood that random initialization of weights followed by gradient descent will lead to good solutions.

- **Stochasticity Helps Escape Saddles:** The stochastic nature of SGD, due to mini-batch sampling, injects noise into the parameter updates. This noise can help the algorithm escape saddle points by perturbing the parameters in random directions.

- Effective Learning Rates: Adaptive learning rate methods (e.g., Adam, RMSprop) adjust the step size based on gradient magnitudes, which can improve convergence near saddle points and in flat regions. In practice, practitioners often use Adam to identify a good region as a starting model and then switch to SGD for further improvement.

- Regularization Effects of Over-parameterization: Over-parameterized networks (with more parameters than training samples) have redundant degrees of freedom, which may facilitate finding flat minima that generalize well.

Current research on trainability and optimization landscapes reveals that, despite the non-convexity of neural network loss functions, certain properties of these functions in high-dimensional spaces make them amenable to optimization via gradient-based methods. The analogy to spin glass models provides a theoretical framework for understanding the distribution of local minima, while the analysis of deep linear networks offers concrete results about the absence of poor local minima and the nature of saddle points.

These theoretical insights help explain the empirical success of training deep neural networks and lay the groundwork for further research into optimization algorithms and network architectures. These will be helpful perspectives to improve training efficiency and generalization performance of deep neural networks.

6.2 The Paradox of Generalization

Despite our discussion on the expressive power and trainability of deep neural networks, a fundamental question still remains unanswered: *Why do these highly overparameterized models generalize well to unseen data?* This phenomenon is often referred to as the paradox of generalization in deep learning and the associated double-descent phenomenon. Classical statistical learning theory, which relies on concepts such as the VC dimension and Rademacher complexity, predicts that models with a capacity far exceeding the number of training samples should overfit the data and exhibit poor generalization performance. However, empirical evidence contradicts this prediction, as deep neural networks often achieve a low generalization error even when they are capable of perfectly fitting random labels.

6.2.1 Empirical observations of over-parameterization

Zhang et al. conducted a series of experiments demonstrating that deep neural networks can achieve zero training error on datasets with randomly assigned labels. Specifically, let $\mathcal{D} = \{(x_i, y_i)\}_{i=1}^{N}$ be a dataset where $x_i \in \mathbb{R}^d$ are input features and $y_i \in \mathcal{Y}$ are labels drawn independently from a random distribution over the label space $\mathcal{Y}$. A neural network $f_\theta : \mathbb{R}^d \to \mathcal{Y}$ with parameters θ can be trained to minimize the empirical risk of the training set:

$$\hat{R}(\theta) = \frac{1}{N} \sum_{i=1}^{N} \ell(f_\theta(x_i), y_i),$$

where $\ell : \mathcal{Y} \times \mathcal{Y} \to \mathbb{R}_+$ is a loss function (e.g., cross-entropy loss). Despite the randomness of the labels, the network can achieve zero training error with $\hat{R}(\theta) = 0$, indicating perfect interpolation of the training data.

However, when trained on real datasets with true labels, the same networks not only achieve low training error but also exhibit low test error on unseen data. This dichotomy suggests that the capacity of neural networks alone does not fully explain their generalization behavior.

6.2.2 Limitations of classical capacity measures

Classical learning theory uses capacity measures such as the VC dimension and Rademacher complexity to provide upper bounds on the generalization error, which quantifies the true risk of a given classifier f_θ and is typically unobserved as the underlying data-generating process $\mathcal{P}$ is unavailable. For a hypothesis class $\mathcal{H}$ and a loss function ℓ, the generalization error $R(\theta)$ of a model $f_\theta \in \mathcal{H}$ is defined as:

$$R(\theta) = \mathbb{E}_{(x,y) \sim \mathcal{P}}[\ell(f_\theta(x), y)],$$

where $\mathcal{P}$ is the underlying data distribution. The goal of learning is thus to bound the generalization gap between $R(\theta)$ and $\hat{R}(\theta)$, where the latter is calculated using the training set and is referred to as the empirical risk.

6.2.2.1 VC dimension

As a recap of Chapter 2, the VC dimension $\text{VC}(\mathcal{H})$ is a measure of the capacity of the hypothesis class $\mathcal{H}$. It is defined as the largest integer h such that there exists a set of h points that can be shattered by $\mathcal{H}$. A set is shattered if, for every possible assignment of labels to the points, there exists a particular hypothesis (a model) in $\mathcal{H}$ that perfectly classifies these points to the correct labels.

For binary classification with 0-1 loss, classical bounds state that, with probability at least $1 - \delta$:

$$R(\theta) \le \hat{R}(\theta) + \sqrt{\frac{\text{VC}(\mathcal{H})(\ln(2N/\text{VC}(\mathcal{H})) + 1) - \ln(\delta/4)}{N}},$$

where N is the number of training samples.

In deep neural networks, the VC dimension can be very large. For example, for a network with W parameters and threshold (binary step) activation functions, the VC dimension satisfies:

$$VC(\mathcal{H}) = O(W \ln W).$$

When $W \gg N$, the bound becomes vacuous, thus providing no useful information about the generalization behavior of the neural network.

6.2.2.2   Rademacher complexity

The empirical Rademacher complexity $\hat{\mathfrak{R}}_N(\mathcal{H})$ of a hypothesis class $\mathcal{H}$ is another capacity measure defined as:

$$\hat{\mathfrak{R}}_N(\mathcal{H}) = \mathbb{E}_\sigma \left[\sup_{h \in \mathcal{H}} \frac{1}{N} \sum_{i=1}^{N} \sigma_i h(x_i) \right],$$

where σ_i are independent Rademacher variables that take values ± 1 with equal probability. The Rademacher complexity thus measures how well the class $\mathcal{H}$ can fit random noise.

For models with high capacity, such as deep neural networks with many parameters, the Rademacher complexity can be large, leading to loose generalization bounds. Specifically, the generalization error can be bounded by:

$$R(\theta) \le \hat{R}(\theta) + 2\hat{\mathfrak{R}}_N(\mathcal{H}) + 3\sqrt{\frac{\ln(2/\delta)}{2N}}.$$

Again, when $\hat{\mathfrak{R}}_N(\mathcal{H})$ is large, the bound may not be informative.

6.2.3   Explaining the paradox

The failure of classical capacity measures to characterize the generalization behavior of deep neural networks has led researchers to explore alternative explanations, such as using implicit regularization, neural tangent kernel, and the double-descent phenomenon, among other ongoing directions. Let us review these perspectives in more detail.

6.2.3.1   Implicit regularization

One hypothesis is that the training algorithm itself could impose an implicit regularization effect that biases the solution toward models that tend to generalize well. For example, SGD preferentially looks for solutions with certain properties (e.g., minimal norm or smoothness) that are not directly captured if we directly minimize the empirical loss of the training set and search for the optimal model in the hypothesis class.

Consider linear regression with minimum ℓ_2 norm of the prediction error. Given an underdetermined system $Xw = y$, where $X \in \mathbb{R}^{N \times d}$ with $N < d$, the minimum-norm solution is given by:

$$w^* = X^\top (XX^\top)^{-1} y.$$

Gradient descent initialized at $w = 0$ could still converge to w^*, even though there are infinitely many solutions that fit the data perfectly. This suggests that the optimization dynamics can be selected among multiple interpolating solutions in a way that affects the generalization behavior of the trained model.

6.2.3.2   Neural tangent kernel (NTK) perspective

The NTK framework, as discussed in Jacot et al. (2018), models the training dynamics of neural networks in a limiting paradigm of infinite width. In this regime, the network behaves like a kernel method with a specific kernel determined by the prespecified network architecture and initialization method. The NTK remains constant during training, and the evolution of the output function $f_t(x)$ follows linear dynamics:

$$f_t(x) = f_0(x) - \eta\Theta(x, X)(I - e^{-\eta\Theta(X,X)t})(f_0(X) - y),$$

where η is the learning rate, Θ is the NTK, f_0 is the initial network prediction, and t denotes time.

This perspective suggests that even highly nonlinear networks can be approximated by linear models during training, and generalization can be analyzed using kernel methods. However, the NTK approach may not fully capture the behavior of finite-width networks, which is the mainstream in practical architectural design.

6.2.3.3   Double-descent phenomenon

Belkin et al. (2019) observed that the generalization error, when plotted as a function of model capacity, can exhibit a double-descent behavior. In particular, increasing the model capacity initially leads to a lower training error and an increased test error, following the classical overfitting regime of a U-shaped curve. However, beyond a certain point (the interpolation threshold at which the training error reaches zero), further increasing the capacity leads to a decrease in the test error.

Formally, let $\mathcal{H}_\alpha$ denote a hypothesis class parameterized by capacity α. The test error $R(\theta_\alpha)$ can exhibit a non-monotonic behavior as follows:

$$R(\theta_\alpha) = \begin{cases} \text{Decreases} & \text{for } \alpha \leq \alpha_c, \\ \text{Increases} & \text{for } \alpha_c < \alpha < \alpha^*, \\ \text{Decreases} & \text{for } \alpha \geq \alpha^*, \end{cases}$$

where α_c is the "sweet spot" with the lowest test error in the classical U-shaped curve, and α^* is the minimal training error (the interpolation threshold), after which the test error starts to decrease again.

This phenomenon challenges the traditional view on bias-variance trade-off and suggests that over-parameterization can, counterintuitively, improve the generalization performance of many overly complex models.

6.2.4 Alternative theoretical frameworks

To address the paradox of generalization, researchers have also proposed alternative theoretical frameworks that consider factors beyond the capacity of the hypothesis class. These include margin-based analysis, algorithmic stability, and compression-based approach.

6.2.4.1 Margin-based analysis

Margin theory, originally developed for support vector machines, has been extended to neural networks. The margin of a classifier on a sample (x_i, y_i) is defined as:

$$\gamma_i = y_i f_\theta(x_i),$$

assuming binary labels $y_i \in \{-1, +1\}$.

The empirical margin distribution can provide insights into the generalization behavior. For example, Neyshabur et al. (2015) derived generalization bounds based on the norms of the weight matrices and the margin:

$$R(\theta) \le \hat{R}_\gamma(\theta) + O\left(\frac{\prod_{l=1}^{L} \|W^{(l)}\|_F}{\gamma \sqrt{N}} \right),$$

where $\hat{R}_\gamma(\theta)$ is the empirical margin loss, $\|W^{(l)}\|_F$ is the Frobenius norm of the weight matrix at layer l, and L is the number of layers. Such bound differs from previous ones in that the true risk is now bounded by the empirical margin loss, plus a constant term proportional to the norm of weight matrices.

6.2.4.2 Algorithmic stability

Algorithmic stability measures how sensitive the output of a learning algorithm is to changes in the training data. An algorithm is uniformly stable if, for all datasets S and S' differing by one sample, and for all inputs x:

$$|\ell(f_S(x), y) - \ell(f_{S'}(x), y)| \le \beta,$$

where f_S and $f_{S'}$ are the models trained on S and S', respectively.

Hardt et al. (2016) showed that SGD has algorithmic stability properties that can be used to bound the generalization error:

$$\mathbb{E}[R(\theta) - \hat{R}(\theta)] \le \beta.$$

This suggests that the learning dynamics, when using SGD to optimize model parameters, contribute to generalization by limiting the model's sensitivity to individual training samples.

6.2.4.3 Compression-based approaches

Compression schemes attempt to explain generalization by showing that a trained network can be compressed to a smaller network or a simpler representation without significant loss in performance. The idea is that if the learned function can be described succinctly, then it effectively belongs to a smaller hypothesis class with a lower capacity.

In particular, Arora et al. (2018) developed a compression-based generalization bound:

$$R(\theta) \leq \hat{R}(\theta) + O\left(\frac{\log(\text{size of compressed model})}{N}\right).$$

This approach essentially connects generalization to the information-theoretic complexity of the learned model.

6.3 Open Questions and Future Directions

Despite significant advances in understanding the expressivity and trainability of deep neural networks, the paradox of generalization remains only partially understood. In this section, we introduce a few key open questions that continue to challenge researchers and outline potential directions for future work.

6.3.1 Role of data distribution

One of the critical aspects influencing generalization is the nature of the underlying data distribution. Real-world data often possess inherent structures, such as low-dimensional manifolds embedded in high-dimensional spaces, sparsity (only a few features are significant out of many), or hierarchical patterns (high-level features are composed of low-level features). Understanding how these properties contribute to generalization is a vital open question.

For example, the manifold hypothesis posits that high-dimensional data lie approximately on a manifold of much lower dimensionality. Mathematically, let $\mathcal{M} \subset \mathbb{R}^D$ be a manifold of dimension $d \ll D$, and data samples $x_i \in \mathcal{M}$. Neural networks can exploit this low-dimensional structure to generalize effectively.

Analyzing generalization in the context of the manifold hypothesis involves studying how neural networks interpolate between data points on the manifold and how they extrapolate off the manifold. Techniques from differential geometry and manifold learning could provide theoretical frameworks to quantify the impact of data geometry on generalization.

Traditional capacity measures, such as those mentioned previously, often do not account for the underlying distribution of the data. Developing data-dependent complexity measures, such as covering numbers or mutual information between inputs and outputs, can lead to tighter generalization bounds.

For example, suppose that the data distribution $\mathcal{P}_X$ has support on a set with covering number $\mathcal{N}(\varepsilon)$ at scale ε. Generalization bounds can be refined to:

$$R(\theta) \leq \hat{R}(\theta) + O\left(\sqrt{\frac{\ln \mathcal{N}(\varepsilon)}{N}}\right),$$

where N is the number of training samples. Understanding how properties such as intrinsic dimension, clustering, and margin distribution influence the covering number $\mathcal{N}(\varepsilon)$ is an open area of research.

We outline three potential directions for future research related to the underlying data-generating process:

- Developing theories that incorporate data geometry and structure into generalization analyses.

- Investigating how data augmentation and transformations affect the effective data distribution and generalization.

- Studying generalization in settings with non-IID data, such as time series or adversarial environments.

6.3.2 Implicit bias of optimization algorithms

The optimization algorithm used to train neural networks plays a crucial role in determining which solution is found among the many that fit the training data. The implicit bias or implicit regularization of algorithms like SGD may guide the search toward solutions with favorable generalization properties.

For example, in linear models, it is well understood that gradient descent converges to the minimum norm solution. For example, in linear regression, gradient descent initialized at zero converges to the solution:

$$w^* = \arg\min_{w} \|w\|_2 \quad \text{subject to} \quad Xw = y.$$

This phenomenon suggests that gradient descent implicitly regularizes the search by favoring solutions with smaller norms.

However, directly extending these results to deep, nonlinear networks is challenging. Recent theoretical work attempts to characterize the implicit bias of gradient-based optimization in deep learning. For example, Soudry et al. showed that for linearly separable data and logistic regression, gradient descent converges in direction to the maximum-margin separator. In deep networks, researchers have proposed that SGD biases the network toward solutions with certain norm constraints or low complexity.

Mathematically, suppose that we define a complexity measure $\Omega(\theta)$ (e.g., weight norms). The implicit bias hypothesis suggests that SGD solves:

$$\theta^* = \arg\min_{\theta} \Omega(\theta) \quad \text{subject to} \quad \hat{R}(\theta) = 0.$$

Understanding what $\Omega(\theta)$ represents in deep networks and how it affects generalization still remains an open question.

In general, analyzing the stochastic dynamics of SGD, including the effects of learning rate, batch size, and momentum, can shed light on the implicit bias. Some future directions of research include:

- Studying the stationary distribution of SGD and its relation to flat minima.

- Exploring connections between SGD dynamics and Bayesian inference, where SGD approximates sampling from a posterior distribution.

- Developing theoretical models that capture the implicit bias of optimization algorithms in deep networks.

- Investigating how different optimization hyperparameters influence the bias and generalization.

- Extending implicit bias analyses to other optimization methods, such as adaptive algorithms (e.g., Adam, RMSprop).

6.3.3   Bridging theory and practice

There is often a gap between theoretical models and practical observations in deep learning. Bridging this gap is essential for developing theories that can guide the design of better model architectures and optimization algorithms.

There are quite some phenomena that are observed but not fully explained by current theories. A few examples include:

- The effectiveness of specific architectures (e.g., ResNets, Transformers).

- The impact of training techniques such as dropout, batch normalization, and data augmentation.

- The role of hyperparameters in training stability and generalization.

One future direction of research on this front is to create benchmarks and standardized datasets for theoretical investigation. Potential research directions also extend to:

- High-Dimensional Probability: Techniques for analyzing random processes in high-dimensional spaces.

- Non-Convex Optimization: Advances in understanding global convergence properties in non-convex settings based on frameworks such as Bayesian optimization Liu (2023).

- Information Theory: Applying information-theoretic concepts to quantify generalization and compression.

- Statistical Physics: Utilizing analogies with physical systems to model learning dynamics and loss landscapes.

6.4 Summary

In this chapter, we explored the foundational theories and contemporary challenges associated with understanding generalization in deep learning. The remarkable empirical success of deep neural networks in various complex tasks has prompted a deeper investigation into the theoretical principles that underpin their performance, particularly concerning their ability to generalize from training data to unseen test samples.

We began by examining the expressivity of neural networks, highlighting the Universal Approximation Theorem, which says that feedforward networks with a single hidden layer can approximate any continuous function on a compact subset of $\mathbb{R}^n$, given sufficient capacity. This theorem underscores the potential of neural networks to model complex functions but leaves open questions about the efficiency of approximation and the practical limitations in terms of network size and computational resources.

We then introduced approximation rates and the trade-offs between depth and width in network architectures. For example, Barron's approximation bounds provided quantitative insights into how the approximation error decreases with the number of neurons, while Telgarsky's results emphasized the exponential advantages that a deep neural network can excel in representing certain functions. These findings inform the design of neural networks by illustrating how depth and width contribute differently to a network's expressive power.

The chapter also addressed the trainability of neural networks by exploring the optimization landscapes associated with training deep models. Despite the non-convexity of the loss functions, which theoretically could hinder optimization due to multiple local minima and saddle points, empirical evidence shows that gradient-based methods like SGD are effective. We discussed the spin glass analogy and the properties of deep linear networks to understand why, in high-dimensional parameter spaces, most local minima are nearly as good as the global minimum, and saddle points rather than poor local minima pose the primary challenge.

Central to our discussion was the paradox of generalization. Classical statistical learning theories, relying on capacity measures such as the VC dimension and Rademacher complexity, seemingly fail to explain why overparameterized neural networks generalize well. These theories predict that models with capacity exceeding the number of training samples should overfit, yet deep networks deviate from this expectation. We discussed empirical observations that demonstrate the fact that neural networks can perfectly fit random labels but still generalize effectively on real data.

To address this paradox, we explored alternative theoretical frameworks:

- Implicit Regularization: The idea that the optimization algorithms themselves introduce a bias toward solutions that generalize well, even in the absence of explicit regularization terms.

- Neural Tangent Kernel (NTK): A perspective where, in the infinite-width limit, neural networks behave like kernel methods, allowing for the application of kernel theory to analyze training dynamics and generalization.

- Double-Descent Phenomenon: An observed behavior where increasing model capacity beyond a certain point leads to improved generalization, challenging the traditional bias-variance trade-off.

Finally, we identified open questions and future directions that remain at the forefront of research on generalization in deep learning models:

- Understanding the role of data distribution, including how real-world data structures like low-dimensional manifolds impact generalization.

- Deciphering the implicit bias of optimization algorithms and how methods like SGD influence the selection of solutions in the parameter space.

- Investigating the conditions under which over-parameterization benefits generalization and how it relates to the geometry of the loss landscape.

- Bridging the gap between theory and practice by developing theoretical models that accurately map to empirical observations in deep learning.

Addressing these questions necessitates a multidisciplinary approach, combining theoretical innovation with empirical validation, and may involve the development of new mathematical tools capable of capturing the complexities inherent in modern neural networks.

Building upon the theoretical foundations and open questions presented in this volume, Volume II (Next volume of the book (to be written next year)) of this book will dive into an in-depth exploration of regularization in deep learning. Regularization is a crucial aspect of training neural networks, aimed at improving generalization performance by preventing overfitting to the training data. Specifically, we will cover commonly used regularization techniques through four primary lenses: data, objective/cost function, model hyperparameter configuration, and optimization algorithm. Throughout Volume II, we will integrate theoretical analyses with practical implementations, providing insights into why these regularization techniques work and how they can be effectively applied in various deep learning contexts. By systematically examining regularization from these multiple perspectives, we aim to equip readers with a comprehensive understanding of the strategies that contribute to building models that not only fit the training data well but also generalize effectively to new, unseen scenarios.

Thank you for taking the time to read this book up to this point. I look forward to seeing you again in Volume II!

Bibliography

[1] S. Arora, R. Ge, B. Neyshabur, and Y. Zhang. Stronger generalization bounds for deep nets via a compression approach. In J. Dy and A. Krause, editors, *Proceedings of the 35th International Conference on Machine Learning*, volume 80 of *Proceedings of Machine Learning Research*, pages 254–263, Stockholmsmässan, Stockholm, Sweden, Jul 10–15 2018.

[2] A. R. Barron. Universal approximation bounds for superpositions of a sigmoidal function. *IEEE Transactions on Information Theory*, 39(3):930–945, 1993.

[3] M. Belkin, D. Hsu, S. Ma, and S. Mandal. Reconciling modern machine-learning practice and the classical bias variance trade-off. *Proceedings of the National Academy of Sciences*, 116(32):15849–15854, 2019.

[4] A. Choromanska, M. Henaff, M. Mathieu, G. Ben Arous, and Y. LeCun. The loss surfaces of multilayer networks. In G. Lebanon and S. V. N. Vishwanathan, editors, *Proceedings of the Eighteenth International Conference on Artificial Intelligence and Statistics*, volume 38 of *Proceedings of Machine Learning Research*, pages 192–204, San Diego, CA, USA, May 9–12 2015. PMLR.

[5] Y. N. Dauphin, R. Pascanu, Gülçehre, K. Cho, S. Ganguli, and Y. Bengio. Identifying and attacking the saddle point problem in high-dimensional non-convex optimization. *arXiv preprint arXiv:1406.2572*, 2014.

[6] M. Hardt, B. Recht, and Y. Singer. Train faster, generalize better: Stability of stochastic gradient descent. In M.-F. Balcan and K. Q. Weinberger, editors, *Proceedings of the 33rd International Conference on Machine Learning*, volume 48 of *Proceedings of Machine Learning Research*, pages 1225–1234, New York, NY, USA, June 20–22 2016. PMLR.

[7] A. Jacot, F. Gabriel, and C. Hongler. Neural tangent kernel: Convergence and generalization in neural networks. In *Advances in Neural Information Processing Systems*, 31:8571–8580, 2018.

[8] K. Kawaguchi. Deep learning without poor local minima. In *Advances in Neural Information Processing Systems*, 29:586–594, 2016.

[9] P. Liu. *Bayesian Optimization: Theory and Practice Using Python*. Apress, 2023.

[10] B. Neyshabur, R. Tomioka, and N. Srebro. Norm-based capacity control in neural networks. In P. Graz, U. von Luxburg, and R. Garnett, editors, *Proceedings of the 28th Annual Conference on Learning Theory*, volume 40 of *Proceedings of Machine Learning Research*, pages 1376–1401, Paris, France, July 3–6 2015. PMLR.

[11] D. Soudry, E. Hoffer, M. S. Nacson, S. Gunasekar, and N. Srebro. The implicit bias of gradient descent on separable data. *The Journal of Machine Learning Research*, 19(1):2822–2878, 2018.

[12] M. Telgarsky. Benefits of depth in neural networks. In V. Feldman, A. Rakhlin, and O. Shamir, editors, *Proceedings of the 29th Annual Conference on Learning Theory*, volume 49 of *Proceedings of Machine Learning Research*, pages 1517–1539, Columbia University, New York, NY, USA, June 23–26 2016. PMLR.

[13] C. Zhang, S. Bengio, M. Hardt, B. Recht, and O. Vinyals. Understanding deep learning requires rethinking generalization. *arXiv preprint arXiv:1611.03530*, 2017.

[14] Preetum Nakkiran, Gal Kaplun, Yamini Bansal, Tristan Yang, Boaz Barak, and Ilya Sutskever (2000). Deep suble descent: Where bigger models and more data hurt. In ICLR 2020, https://dblp.org/rec/bib/journals/corr/abs-1912-02292

[15] Samuel L. Smith, and Benoit Dherin, and David G. T. Barrett (2021). On the origin of implicit regularization in stochastic gradient descent. Accepted at ICLR 2021. https://arxiv.org/abs/2101.12176

Index